T0177599

Ranking

Ranking

The Unwritten Rules of the Social Game We All Play

PÉTER ÉRDI

Kalamazoo College

Wigner Research Centre for Physics of the Hungarian Academy of Sciences, Budapest

Oxford University Press is a department of the University of Oxford. It furthers the University's objective of excellence in research, scholarship, and education by publishing worldwide. Oxford is a registered trade mark of Oxford University Press in the UK and certain other countries.

Published in the United States of America by Oxford University Press
198 Madison Avenue, New York, NY 10016, United States of America.

CIP data is on file at the Library of Congress
ISBN 978-0-19-093546-7

1 3 5 7 9 8 6 4 2

Printed by Sheridan Books, Inc., United States of America

For my children, Gábor and Zsuzsi

Contents

Contents

Foreword

In this brilliant, wide-ranging book, Péter Érdi, an award-winning teacher and scholar, takes up the phenomena of rankings and ratings (I'll get to the difference in a moment). A computational scientist, Érdi proves equally skilled as a social observer, revealing deep implications of the ubiquitous rankings and ratings created by social and mainstream media. We assign far too much credibility to numerical rankings that may, at their core, be subjective impressions. Even more troubling, as we change our behavior to move up those lists, we allow ourselves to be manipulated by the rankings.

Rankings and these behavioral responses to them occur across the socio-technical landscape. To grasp the breadth of his inquiry, take a moment and leaf through the index. Here's one sampling taken back to front: the Wong-Baker pain scale, *U.S. News & World Report* university rankings, Scar (yes, from *The Lion King*), recommendation letters, the illusion of objectivity, the Hungarian national soccer team, Erdős numbers, Elo chess ratings, and Campbell's law. The A's alone include Jane Austen, Aristotle, and Arrow's impossibility theorem.

That sampling of topics only hints at the fun in store for you. Throughout, Péter's fertile, lively mind is on full display. You are in for a treat. Though the book takes up technical topics, Péter's writing manages to be bright, funny, and clear. By book's end, many readers may contemplate catching the train to Kalamazoo or Budapest with the hopes of meeting up with Péter to learn more about preferential attachment mechanisms, bounded rationality, social neuroscience, the psychology of list making, or the applications of network statistics. For those of you who know Péter, particularly his former students, reading the book will remind you of his boundless, generous curiosity. The book, like Péter, is informative, deep, thought-provoking, and joyful.

The best rankings rely on objective criteria. Rankings of the tallest buildings, largest Northern Pike, and fastest motorcycles can be accepted at face value. However, even objective criteria may, when viewed under a microscope, contain elements of subjectivity. The official height of a building includes the building's towers if they are integral to the building. The spire on the Freedom Tower in New York counts, while the two antennae atop the Willis Tower in Chicago do not. The integral portion of a building lies in the eye of the beholder. And that's where the problems start, with the inclusion of subjectivity.

Subjectivity allows us to rank as we see fit. In the movie version of Thomas Wolfe's *The Right Stuff*, a reporter asks the astronaut Gordon Cooper, played by a young Dennis Quaid, to name the best pilot he ever saw. Cooper first rambles about pictures on a wall in a place that no longer exists and hurtling steel, all as a prelude to naming Chuck Yeager. When Cooper realizes that the press wants a story and has little interest in the truth of the matter, he breaks into a toothy smile and says, "Who's the best pilot I ever saw? Well, uh, you're lookin' at him."

Air & Space Magazine would beg to differ. They do not rank Gordon Cooper in the top 10, though Yeager does rank. Érdi would be quick to point out that both Cooper's ranking and the magazine's, just like the ubiquitous rankings we find on the World Wide Web—the top 10 beaches, the top eight Belgian ales, and the top seven dog breeds—are subjective. Some person, or group of people, made up an ordering and then justified it using criteria derived after the fact. Nevertheless, these rankings confer a degree of authority—ah, the power of numbers!

Yet, as Érdi shows, in most of the important cases, objective rankings are not possible. Permit me a brief foray into formalism. Formally, a ranking is a *complete, asymmetrical*, and *transitive* relation. "Complete" means that it compares any two things. "Asymmetrical" means that it ranks each item either above or below every other: either you like beets more than carrots or carrots more than beets. "Transitive" means that if A is preferred to B, and B to C, then A must also be preferred to C.

As logical as transitivity may seem, it can be violated by collections of rankings. In a *Condorcet triple*, a majority-rule vote of three people, each of whom has transitive preferences, results in A defeating B, B defeating C, and C defeating A. Majority-rule voting becomes an instantiation of the rock, paper, scissors game. In other words, even if each person has a consistent ranking, that in no way implies that a collective ranking exists.

A similar problem arises if the items we wish to rank possess multiple dimensions. Magazines rank restaurants by evaluating the quality of their food, their ambiance, and the professionalism of their staff. They then assign numbers to each restaurant on each dimension and sum them to produce a rating. Out of a total score of 30, one restaurant may score 28, while another scores 27. These numbers, as Érdi points out, are subjective. One person's five out of five may be another person's four out of five. What appears scientific is largely made up.

In fairness, often the scores are a mix of objective and subjective. Such is the case in the *U.S. News & World Report* college rankings, which take into account the number of classes a college offers with fewer than 19 students and the overall faculty-to-student ratio (both objective) as well as ranking by deans (subjective). To create ratings (which can then be simplified to a ranking), *U.S. News* then attaches a weight to each of these criteria. How do they come up with the weights? Again, those are just made up based on common sense. So once again, what appears scientific is in fact subjective.

An immediate consequence of this method is that a college can improve its ranking by limiting enrollments in some classes to only 19 students. Doing so improves the school's ranking. Keep in mind that no empirical evidence supports a significant loss in learning from adding a 20th student; *U.S. News* just choose the number 19. To see the pernicious effects of just this one ranking criterion, go to almost any college webpage and you will see that they advertise the number of classes with 19 or fewer students. Colleges prevent students from taking classes (sorry, you're number 20) so as not to hurt their rankings.

It follows, paradoxically, that even well-intentioned attempts to identify the best of us may bring out the worst in some of us, as we distort ourselves to improve our rankings. Thus, the greater importance we attach to these mostly subjective rankings, the more we produce behavioral distortions. With this book, Professor Érdi has done us a service. He has taught us to think more deeply, yet done so with captivating examples and a light hand.

Scott E. Page
Leonid Hurwicz Collegiate Professor of Complex Systems, Political Science, and Economics, University of Michigan-Ann Arbor; External Faculty, The Santa Fe Institute

Preface

This book, of course, is about ranking. Like it or not, ranking is with us. Everybody with whom I have talked in the last two years has seemed to agree that the topic is hot. We are in a paradoxical relationship with ranking: ranking is good because it is informative and objective; ranking is bad because it is biased and subjective and, occasionally, even manipulated. This book is intended to help readers understand the paradoxical nature of ranking procedures, and it offers strategies for coping with this paradox. Ranking begins with comparisons. We like to compare ourselves with others and determine who is stronger, richer, better, or cleverer. Our love of comparisons has led to our passion for ranking. Ranking is about becoming more organized, and we like the idea of being more organized!

Humans are not the only ones to employ ranking; it is the result of an evolutionary process. The concept of "pecking order" among chickens was observed about a hundred years ago, and research has proven that chickens, living together in the same poultry run, organize themselves into a social hierarchy. Social ranking in human societies evolved from the world of animals. This book discusses the "whys" and "hows" of our love and fear of ranking and being ranked through real-life examples, examined from three different angles—reality, illusion, and manipulation—of objectivity.

Ranking applies scientific theories to everyday experience by raising and answering such questions as: Are college ranking lists objective? How do we rank and rate countries based on their fragility, level of corruption, or even happiness? How do we find the most relevant webpages? How do we rank employees? Since we permanently rank ourselves and others, and are also being ranked, the message is twofold: how to prepare the most objective ranking possible and how to accept that ranking is not necessarily reflective of our real values and achievements.

While the book takes examples from social psychology, political science, and computer science, it is not a book for the scientists only. The book is offered to people whose neighbor has a fancier car; employees who are being ranked by their supervisors; managers who are involved in ranking but may have qualms about the process; businesspeople interested in creating better visibility for their companies; scientists, writers, artists, and other competitors who would like to see themselves at the top of a success list; college students who are just preparing to enter the new phase of social competition and believe that the only game in town is maximizing their grade-point average at any expense; computer scientists who design algorithms for recommending products for individuals based on their habits; and people who get unsought recommendations (all of us).

Excellent books have already been published discussing different specific aspects of ranking, varying from mathematical algorithms to ranking academic institutions, countries, political candidates, or websites. *Who's #1?: The Science of Rating and Ranking*, by the mathematicians Amy N. Langville and Carl D. Meyer (Princeton University Press, 2012), was developed from their studies analyzing the Web and offers a broad overview of the mathematical algorithms and methods used to rate and rank sports teams, political candidates, products, Web pages, etc., and it can be found on a shelf for math books. I use the spirit of this book to explain the attempt to making rankings objective and also will show the difficulties of being objective.

The next two books on my list deal with college and university rankings. *Engines of Anxiety: Academic Rankings, Reputation, and Accountability* (Russel Sage, 2016) by two sociologists, Wendy Nelson Espeland and Michael Sauder, analyzes the history and present practice of evaluating and ranking the quality of institutions of higher education, particularly law schools. Ranking not only reflects the past but also forms the future, as the key stakeholders (students, parents, admission offices, administrators) react to a ranked list. This book demonstrates the nature of our paradoxical attitude toward ranking: quantifications of performance are both necessary and a source of anxiety. Ellen Hazelkorn, a leading expert in global higher education, wrote *Rankings and the Reshaping of Higher Education: The Battle for World-Class Excellence* (Palgrave Macmillan, 2016, 2nd ed.), which offers a comprehensive study of education rankings from a global perspective.

Ranking the World: Grading States as a Tool of Global Governance (Cambridge University Press, 2015), edited by Alexander Cooley and Jack Snyder, describes controversial emotions toward ranking countries. International rankings of country performance are characterized by about a hundred different indices, from the "Human Freedom Index" to the "Corruption Perceptions Index" to "World Happiness." A recurring theme is that ranking organizations are not totally independent, and even though some ranked countries (say, China and Russia) occasionally react angrily to performance analyses, they are still interested in the results. This book helped me to come to the conclusion, among others, that the happiest countries in the world also pay a lot in taxes.

Majority Judgment: Measuring, Ranking, and Electing by Michel Balinski and Rida Laraki (MIT Press, 2011) is about ranking political candidates, and the authors argue that "the intent of this book is to show why the majority judgment is superior to any known method of voting and to any known method of judging competitions."

Gundi Gabrielle's *SEO—The Sassy Way to Ranking #1 in Google—When You Have NO CLUE!: A Beginner's Guide to Search Engine Optimization* (Amazon Digital Services, 2017) explains the tricks of pushing your website, blogs, etc. to the top without being penalized by the Google or other Internet authorities.

Two recent books have some overlap with my own goals, and we might have an overlapping readership too. Gloria Origgi's *Reputation: What It Is and Why It Matters* (Princeton University Press, 2017) reviews the contribution of some ranking systems to the formation of reputation from the perspective of an "experimental philosopher." Jerry Z. Muller's *The Tyranny of Metrics* (Princeton University Press, 2018) emerged from his observation that measuring and quantifying human performance has too much role in the organization of our society. The historian points to the difficulties of navigating between subjective evaluation and objective measurement, and he might have a somewhat different attitude than I hope to represent.

The challenge I faced when beginning this book was to write a popular, easily readable, integrative book on ranking and rating to help the reader understand the rules of the ranking game we all play each day. The main motivation to write this book came from my former assistant and my true best friend, Judit Szente. Since I told her many times that I felt myself a writer who would be able to write for a broad readership, she and her husband, Bart van der Holst, gave me a birthday gift to study writing: they made me enroll at Gotham Writers' Workshop in New York City. I took wonderful classes taught by Roseanne Wells, Francis Flaherty, Cullen Thomas, Kelly Caldwell, and J. L. Stermer.

I am grateful to the community of Kalamazoo College, and specifically to my close colleagues, who provide me with a friendly, intellectual atmosphere. I am also indebted to my colleagues at the Department of Computational Sciences at Wigner Research Centre for Physics of the Hungarian Academy of Sciences in Budapest. I thank the Henry R. Luce Foundation for letting me serve as a Henry R. Luce Professor.

Natalie Thompson, a political science and math double major, has been serving as my assistant. She not only copy-edited the original "Hunglish" version but also gave comments on the drafts of each chapter from the bird's-eye perspective, and she helped in designing the structure of the book. Her help went way beyond my expectations. Thank you, Natalie!

I benefited very much from the interaction of my old boy network in Budapest. Particularly I am thankful to the comments of Peter Bruck, George Kampis, András Schubert, and János Tóth. In the winter term of 2018 I taught a class about the complexity of ranking and interacted with many students. I am particularly grateful for the comments of Allegra Allgeier, Brian Dalluge, Gyeongho Kim, Timothy D. Rutledge, Skyler Norgaard, and Gabrielle Shimko.

I am grateful for comments, conversations, correspondence, and/or moral support from a number of colleagues: Brian Castellani, John Casti, Alexander Cooley, Peter Dougherty, György Fabri, Rabbi Mordechai Haller, István Hargittai, De-Shuang Huang, Bryan D. Jones, Mark Kear, Andrew Mozina, Scott Page, Peter Prescott, Frank Ritter, Eric Staab, András Telcs, Jan Tobochnik, Osaulenko Viacheslav, and Raoul Wadhwa. As I count now, they are from six different countries. I benefited from the questions and comments of a number of lectures I gave in Budapest, Liverpool, and Cambridge (UK). Thank you to János Tőzsér, Zsuzsa Szvetelszky, Károly Takács, De-Shuang Huang, Abir Hussain, and Dhiya Al-Jumeily for the invitations.

I particularly benefited from the comments made on the about-ranking website by Peter Andras, Basabdatta Sen-Bhattacharya, György Bazsa, Zoltán Jakab, Christian Lebiere, András Lőrincz, Ferenc Tátrai, Emanuelle Tognoli, Ichiro Tsuda, and Tamás Vicsek.

I am thankful to my editor at Oxford University Press, Joan Bossert, for her guidance and encouragement.

I have had a long experience with my wife, Csuti, rating and ranking the options of life. I benefited very much from her support, love, and wisdom. It is difficult to express my gratitude.

Péter Érdi
Kalamazoo, Michigan, and Budapest
December 2018

1

Prologue

My early encounters with ranking

How to lead the popularity list? Own a soccer ball!

It is impossible to play soccer if you don't have a ball. But we had one, so we played! I grew up in Budapest (well, in the flat *Pest* and not in the hilly *Buda*, as my wife did, but I promised her I would not make jokes anymore about the cultural differences in the two parts of the city) not long after the war. My elementary school had students (actually boys; no coed schools existed at that time) from Angyalföld (the now-disappearing working class's "Land of Angels") and Újlipótváros ("New Leopold Town," inhabited by middle-class intellectuals of Jewish origin). While there was an obvious social contrast in the backgrounds of our parents (I overlook here the sad family stories hidden by the parents of the New Leopold Town kids), the love of soccer bridged the gap. In the early 1950s, Hungary had the world's best soccer team, led by Ferenc Puskás (1927–2006), whose left foot made him one of the greatest players of all time. This book is about ranking, and I share the opinion of many that he was one of the two best-known Hungarians of the 20th century (Béla Bartók [1881–1945] is arguably the other one). The Hungarian team remained unbeaten for 33 games, a period stretching from 1950 to 1954, ending with a historic loss against West Germany in the 1954 World Cup (the new Germany's first postwar success). I will go back to this story in

Chapter 2, when I investigate the sadness of being ranked "second best." But the point remains that soccer was extremely popular, and almost all of us played nearly every day for eight years.

But in our classroom of 40 boys, our teacher once asked us each to write an answer to the question, "Who is your best friend?" Our answers were anonymous. Thirty-seven votes went to Péter Erdélyi. He had a wonderful sense of humor, but this was not the reason for his big win. His father was a director of a state-owned (what else?) company called "Cultural Articles," dealing with expensive soccer balls. We lived in a poor country, so everything that we could buy in the shops was expensive. So Péter was the *only* boy in the class who had a *real* soccer ball. We *really were* so thankful to have the chance to play with a real soccer ball that we felt Péter was our best friend. No doubt he led the popularity list for the whole year. (I told this story many times during introductory classes on network theory in order to demonstrate star-like organization, as Figure 1.1 shows.)

By this example I intend to illustrate that the selection of the leader of our popularity list objectively reflects *the wisdom of the crowd*, which is neither illusion nor manipulation. As I am meditating on the story, I note that Péter came from a privileged family. To be a very privileged boy in 1950s Budapest meant that he had a soccer ball. The combination of this privileged situation with his nice personality traits pushed him to the top of the popularity list.

Rating and ranking of soccer players: the illusion of objectivity

I must have been only 10, maybe 11, at the time, but I still remember well the paradoxical title of a journal article I once read: "Let the objective numbers speak!" I will enlighten you as to why it was paradoxical. At the end of each soccer season the sports newspaper would evaluate the performance of the players in each of the 11

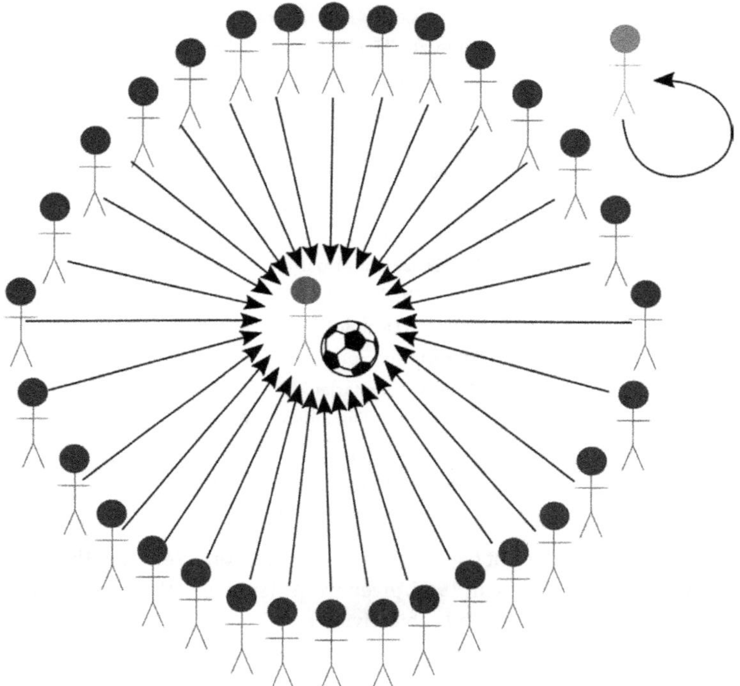

Fig. 1.1 Star-like organization: The kid who had a real soccer ball was the best friend of everybody. Well, everybody except one boy (I know his name but will not mention it, although I will say he has lived in Toronto for many years). Thank you to Tamás Kiss for the figure.

positions, from goalkeepers to left wingers. The article, in addition to verbal appraisal, contained 11 ranked lists, one for each position, and players from each team were ranked based on their seasonal scores (Figure 1.2). How were these scores constructed? Soccer is unlike baseball in that there is no objective measure for scoring the players. (Well, it changed somewhat in the recent years, and a set of performance metrics has been adopted now.) An apprentice journalist was delegated to every game, and he (surely a "he" at the time I am describing) gave a score to each player after each game. Any

Jobbhátvédek

1.	Káposzta (U. Dózsa)	7.13
2.	Bakos (Vasas)	7.06
3.	Hernádi (Pécs)	6.99
4.	Várhelyi III (Szeged)	6.88
5.	Kárpáti (Eger)	6.77
6.	Keglovich (Győr)	6.76
7.	Vellai (Csepel)	6.71
8.	Novák (Ferencváros)	6.68
9.	Lévai (Tatabánya)	6.59
	Kelemen (Komló)	6.59
11.	Marosi (Bp. Honvéd)	6.52
12.	Kmetty (Salgótarján)	6.42
13.	Formaggini (Dunaújv.)	6.41
	Kovács (Diósgyőr)	6.41
15.	Keszei (MTK)	6.30
16.	Szabó B. (Szombathely)	5.51

Fig. 1.2 Ranking of right backs based on their seasonal scores in the Hungarian soccer league in 1967, given by the (subjective) evaluation of journalists and later averaged (objectively).

player who made it onto the field got a score of at least one. A very select few players in each season received a score of 10 for their extraordinary performances. The majority of the scores were in the interval between five and eight, which more or less meant between "somewhat below average" and "excellent (but not brilliant)." After each game, as I walked with my father to the tram stop to get a ride from the suburb called Újpest, where the stadium was located, to our apartment in Újlipótváros, we also gave our own scores to each of the players on our team. After each game, I impatiently awaited the morning paper to compare the journalists' scores with my own. At the end of the season, when I read about "objective numbers," I knew well that they reflected the *objective* average of their *subjective* grades. This observation suggested that ranking based initially on subjective rating generates only the *illusion* of objectivity. The scores were not random—they reflected the best estimates of the journalists—but, beyond dispute, they were subjective.

A not-so-beautiful tale: an example of intentional biased ranking from a Hungarian folktale

László Arany (1844–1898), the son of the celebrated poet and "Shakespeare of ballads" János Arany (1817–1882), collected Hungarian folktales. One of Arany's tales illustrates to children how the strongest participant in a group can manipulate what ought to be a collective decision. The tale proceeds as follow:

> A number of animals escaped from their homes and fell into a trap. They were not able to escape, and they became very hungry. There wasn't any food around, so the wolf in the group suggested a solution: "Well, my dear friends! What to do now? We should eat soon, otherwise we will starve to death. I have an idea! Let us read out the names of all of us, and the most ugly one will be eaten." Everybody agreed. (I have never understood why.) The wolf *assigned himself* to be the judge, and counted: "Woolf-boolf, O! So great! Fox-box also great, my-deer-my-beer very great, rabbit-babbit also great, cock-bock also great, my-hen-my-ben, you are not great," and they ate the hen . . . Next time cock-bock became food, and so on. (Many thanks to Judit Zerkowitz for the translation from Hungarian.)

This example demonstrates on a minor level how objectivity can be manipulated if a single voter controls an election. It also suggests to us a form of despotism, a mechanism of ruling in which one person has total, unchecked power.

Lessons learned: the reality, illusion, and manipulation of objectivity

In the realm of sports, one of the oldest and most objective forms of ranking occurs when runners are judged by their speed, an activity with roots in the ancient Olympic Games in Greece. We know, for example, that Koroibos of Elis, a humble cook by profession, won the *stadion* footrace in the very first Olympic Games, meaning he was the fastest runner in the competition. However, other forms of ranking, like many "top 10" (21, 33, etc.) lists, are based on subjective categorization and give only the *illusion* of objectivity. In fact, we don't necessarily always like objectivity, since we don't mind if our performances, websites, businesses, or organizations have a better image, score, or rank than they deserve. More precisely, occasionally we are the victims of biased self-perception (I'm sure many readers have seen the image of the kitten who looks in the mirror and sees herself a lion[1]), and other times we willfully deceive ourselves and wish to be perceived as having a higher status than we really do. In the latter case, we don't mind manipulating objectivity through a procedure euphemistically called *reputation management*.

Our struggle for reputation will be discussed in Chapter 7. But first, let's review the many concepts to be discussed in this book.

2

Comparison, ranking, rating, and lists

Comparison: The "thief of joy" or the driving force toward future successes?

We constantly compare ourselves with others. In many cultures, children learn they should win competitions to demonstrate that they are better, stronger, and more successful than the others. High school class reunions, for example, provide wonderful opportunities to compare our standing in any aspect of life—from attractiveness to career progress to intelligence to marital success—against the standings of our former classmates. In everyday life, the evaluation of our own attitudes, abilities, and beliefs is based on comparisons with others. This observation constitutes the foundation of a celebrated theory in social psychology, called *social comparison theory*, written about by Leon Festinger (1919–1989) as early as 1954. Although we may not like to see that we are overweight in comparison to our former teammates, generally (well, I wrote "generally," so not always) we have the social skills to control our feelings of envy. Despite the truism "comparison is the thief of joy," attributed to former US president Theodore Roosevelt, we can't help but compare ourselves with others.

Upward and downward comparison

The term *upward (downward) comparison* refers to situations in which a person compares herself with others who are better (worse) than she is. As an example from my own life, as a young man I had two close friends, call them John and Joe. In the 1970s and early 1980s, not everyone in Budapest necessarily owned a car. If they did have a car, it was most likely to be an "Eastern" car, the most common of which was called a Trabant and was produced in East Germany. It used what was called a two-stroke engine, which was obsolete even at that time. It used to be said that two people were needed for its construction—one to cut and one to glue, as it was made from plastic, and many jokes were made about its quality. I recall one in particular that went something like this:

> A donkey and a Trabant meet in the Thuringian Forest.
>
> "Hi, car!" greets the donkey.
>
> "Hi, donkey!" answers the Trabant.
>
> Offended, the donkey replies, "It is not nice to call me donkey if I addressed you as a car. You should have called me at least a horse!"

I bought a six-year-old Trabant in my mid-30s as my first car. It was not a status symbol, but it had four wheels. John did not have a car (both because he couldn't afford one on his salary as a mathematician and because he had high-diopter glasses that prohibited him from obtaining a driver's license). Common knowledge suggests that the positive effect of any downward competition is gratitude, which I certainly felt when comparing myself to the carless John. While I don't believe I felt the textbook negative effect of downward comparison (scorn), I might have experienced some level of superiority. Joe, who worked for a French company, soon got a "Western" car, a Renault type. Did I feel any hope or inspiration, the positive effects of upward comparison that are cited in the textbooks? Perhaps my aspirations increased to being able to afford (well, in a distant future) a "Western" car like Joe. Concerning the negative effects, I cannot deny I was envious. But was John unhappy or frustrated? Absolutely not! Relevance is a necessary condition of social comparison, and he was absolutely not interested in having a car, so he didn't care!

Social psychologists continue to analyze our motivations for comparison, and in their book *Friend and Foe: When to Cooperate, When to Compete, and How to Succeed at Both*, Adam Galinksy (a social psychologist from Columbia University) and Maurice Schweitzer (from the Wharton School of Business at the University of Pennsylvania)[1] write, "When it comes to using social comparison to boost your own motivation, here is the key rule to keep in mind: Seek favorable comparisons if you want to feel happier, and seek unfavorable comparisons if you want to push yourself harder. You may not be able to quit your social-comparison habit, but you can learn to make it work for you."

Making comparisons with your own goals

When I posted some of my preliminary drafts on comparison on my blog, Peter Andras, an open-minded professor of computer science at the University of Keele in England, made the following comment:

> I think that this relates to the extent to which one's decision making style is more externally or internally driven, or the extent of autonomy in the decision making. There is a lot of work

> in the context of education theory and education psychology on the importance of this distinction and the role of autonomy in the development of individuals and their personality. More autonomous persons compare themselves and their achievements and possessions against their own aims. However, in general more externally driven people dominate communities as it had been found in many contexts that comparison with neighbours and others dominate very much the decision making of most people.

Andras's observations are elaborated on in a controversial book entitled *Punished by Rewards*,[2] in which Alfie Kohn argues against the basic strategy we often use to motivate others: "Do this and you will get that." Rewards and punishments are the two sides of manipulating behavior, and authors like Kohn see rewards as specifically damaging, especially if a student, athlete, or employee already has an intrinsic motivation to succeed. Newer data and theories, like those published by Christina Hinton (from the Harvard Graduate School of Education and the founder and executive director of Research Schools International), support the view that motivating students with external rewards, such as money, is insufficient to maintain their interest in learning.[3] However, when students have *intrinsic motives* for learning, they are more likely to become deeply interested in their work, to show more persistence in the face of learning challenges, and to explore and find new topics.

Still, despite these differences, we can't deny that direct comparison is a very important evaluative mechanism in emotionally processing our successes and our fiascoes. A frequently used procedure is *pairwise comparison*, in which a pair of people, objects, or any other entities are judged based on some quantitative properties (say, who is taller, stronger) or qualitative properties (say, preferences or attitudes). Boxing is certainly the paradigm of direct, pairwise comparison.

From Ali's "I am the greatest" to "the grass is always greener"

Superiority versus inferiority complex

Direct comparison can lead to various emotional results, from Muhammad Ali's (1942–2016) famous proclamation "I am the greatest," to the melancholy "The grass is always greener on the other side." Actually, Ali stated even more boldly: "I'm not the greatest. I'm the double greatest. Not only do I knock 'em out, I pick the round. I'm the boldest, the prettiest, the most superior, most scientific, most skillfullest fighter in the ring today." In principle, we may believe that self-qualification is suspicious and leads to biased ranking, but Ali's statement was approved by the "collective wisdom" of the time: almost everyone from the generation who saw him in the ring believes that Ali really was the greatest. Ali's introspection extended elsewhere—when the US Army measured Ali's IQ at 78, he reportedly said, "I only said I was the greatest, not the smartest." I find it amazing how detachedly he described his work: "It's just a job. Grass grows, birds fly, waves pound the sand. I beat people up."

As opposed to the sense of superiority resulting from comparison in Ali's case, another class of comparison can contribute to a perceived inferiority complex. The idea behind the quotation "The grass is always greener on the other side of the fence" may have its origin in the poetry of Ovid (43 BCE–17 or 18 CE), who wrote, in *Art of Love*, "The harvest is always richer in another man's field." There are other proverbs expressing a similar attitude: "The apples on the other side of the wall are the sweetest," "Our neighbor's hen seems a goose," and "Your pot broken seems better than my whole one." The German version of the proverb, "Kirschen in Nachbars Garten schmecken immer besser," loosely translates to "the cherries in the neighbor's garden always taste better." These all convey the message that others have a better life or are more fortunate than we are. The constant feeling that others are better off can lead to a life of misery as envy leads to anxiety and to other mental health problems. The suggestion made by Robert Fulghum, author of the bestselling book *All I Really Need to Know I Learned in Kindergarten*,[4] is not only more objective but also offers a viable strategy: "The grass is not, in fact, always greener on the other side of the fence. No, not at all. Fences have nothing to do with it. The grass is greenest where it is watered. When crossing over fences, carry water with you and tend the grass wherever you are." We need to find the balance between accepting reality and making an effort to change things toward future successes.

Comparison is with us, as we know from history and literature, too, which the next examples illustrate.

Comparison among immigrant groups

The history of immigration in the United States over the last 150 years has been defined by both upward and downward comparisons. New immigrant groups often have seen themselves in competition with African Americans and other racial minorities for available low-wage work. Perhaps the clearest example of this can be seen in the arrival of the Irish in the middle of the 19th century. To establish their right to low-wage work, they often made racist remarks regarding African Americans, practicing downward comparison through the denigration of another group in response to their own disrespect at the hands of White Anglo-Saxon Protestants. Then, when Italian immigrants started to arrive, the Irish snubbed them at their churches, despite the fact that both Irish and Italian immigrants were largely Catholic, to preserve their relative rise in the social pecking order. (The origin of the concept of pecking order will be discussed in Chapter 3.) Throughout history, each new immigrant group has been slotted into the social hierarchy based on perceived stereotypes about the group and its degree of economic clout. These stereotypes are aimed at creating a hierarchy of comparable traits (e.g., intelligence, sobriety, degree of polish, work habit) with which to make upward and downward comparisons and to attempt to justify social stratification.

Comparison in the American literature

I asked my colleague Dr. Andy Mozina, professor of English and accomplished novelist, for a list of what he feels are the best illustrations from American and British literature of the direct effects of comparison. I respect his knowledge and taste, so I decided to quote his suggestions:

- In Jane Austen's novel *Pride and Prejudice*, Darcy, a rich and snobby aristocrat, looks down on Elizabeth Bennet, whose family is noble but has much less money and seems to have cruder manners. The trajectory of the novel, though, is to show that in fact, through their intellect and their characters, they are much more equal than they at first realized.
- In Toni Morrison's *The Bluest Eye*, a Black girl compares her appearance to the cute, blonde, blue-eyed Shirley Temple, seeing how Shirley's looks are valued and hers are not. After receiving a lot of social messages that she is ugly, in part because she's Black, she aspires to have blue eyes to give her some claim to White beauty. She ends up completely losing touch with her intrinsic value as a person and becomes almost 100 percent externally motivated. Her pursuit of blue eyes, combined with the way others treat her, ends up leaving her destroyed as a person.

Social comparison and our brain

Modern neuroscience has adopted the use of brain imaging methods to help identify the brain regions and neural mechanisms responsible for upward and downward comparison.[5] Downward comparison activates a brain region called the ventromedial prefrontal cortex, an area that is also activated in cases like the processing of monetary rewards. Upward comparison correlates with activity in the dorsal anterior cingulate cortex. Interestingly, this region is involved in signaling negative events, such as feeling pain or experiencing a monetary loss. Researchers cautiously suggest that neuropsychological bases of social comparison can be understood in a more general framework of processing rewards and losses, something we have evolved to keep track of. We are specifically sensitive to a comparison when we are just second, and not first.

The tragedy of being the second best

The tragedy (or miracle) of Bern

I can't help it, but I must return to soccer and discuss the final game of the 1954 World Cup, in which Germany (at that time West Germany) won 3–2 against Hungary. I promised to return to this story. The reminiscence of this game is very strong among West German and Hungarian males in my age group. Hungary beat Germany 8–3 in the group stages. However, the team's captain and very best player, the legendary Ferenc Puskás, was injured. Hungary won two wonderful subsequent games against Brazil and Uruguay to reach the final. Although he was not fully fit in time, Ferenc Puskás was back on the Hungarian team for the final match, and he put Hungary ahead after only six minutes. When Zoltán Czibor added the second goal for Hungary a mere two minutes later, everybody believed it would again be an easy victory over Germany. But

Germany caught up quickly, and it won. According to some, "There are several strong indications that point to the injection of pervitin [methamphetamine] in some Germany players and not vitamin C as it was claimed."[6,7] While the West German football team's World Cup win was a real positive turning point in postwar German history, the aftermath in Hungary was a particular illustration of the effect a sporting event can have on a country's politics. The loss was a dramatic shock to the Hungarian public and led to the first spontaneous postwar demonstrations in Budapest, which were directed not only against the football team and its coaches, but also against the whole authoritarian regime in place at the time.

If you are second, you are not the first

Bestselling children's author (and attorney; it makes for an excellent combination) Rachel Renée Russell wrote in *Dork Diaries* in response to the feelings of a female middle school character: "I feel like I'm always second best. I'm always the backup friend, the third wheel. When my teachers tell us to get into pairs, I'm always the one left out. All my friends partner up, and I'm left standing there awkwardly. I'm sick of being everyone's second choice. No matter how hard I try, I'll never be good enough. Please help me!" Russell responds, "So, what if instead of waiting, you picked someone yourself? What if instead of looking dejected, you plastered a big old grin on your face, walked right up to someone before she could choose someone else, and said, 'Want to pair up?"'

Middle school girls are not the only ones with this problem: "Being second is to be the first of the ones who lose," according to Ayrton Senna (1960–1994), arguably the most influential Formula One driver in the sport's history. And silver medalist Abel Kiviat (1892–1991) admitted at the age of 91, "I wake up sometimes and say: 'What the heck happened to me?' It's like a nightmare." Kiviat was supposed to win the 1,500-meter run at the

1912 Olympics in Stockholm when Arnold Jackson "came from nowhere" to beat him by a mere one-tenth of a second.

What happened

While the title of this paragraph is an open allusion to Hillary Clinton's book explaining her loss in the 2016 US presidential election, I remain here on the topic of sports to tell another story of the modern Olympic games. Geza Imre, a legendary Hungarian fencer, won a bronze medal in men's epee individual competition in 1996 in Atlanta. (He did not attend the next Olympic games in Sydney in 2000, so he did not see his wife, Beatrix Kökény, in the final round of the women's handball competition between Denmark and Hungary, in which Denmark grabbed the gold despite Hungary's six-goal advantage at one point during the game.) After winning the world championship in 2015, he was again in the Olympic final in 2016. With the score at 14–10, he was just one touch shy from the necessary 15 to win the gold. His opponent, the South Korean fencer Park Sang-Young, was just one year old when Imre won the bronze in Atlanta. But Park changed the story for Imre: "I was the winner up until eight-and-a-half minutes into the bout and in the last twenty seconds he beat me. The last four touches he changed his tactics and I couldn't do anything," said Imre.

It might be difficult to accept psychologically that somebody is very, very, very close to achieving a big goal, like winning an Olympic championship, and suddenly a failure emerges in the brain, mind, heart, hand, or foot, and the dream, almost fulfilled, is suddenly over. Oft-cited studies in psychology concerning Olympic medalists clearly demonstrate that silver medalists tend to be miserable as a result of comparisons between themselves and the gold medalists. Bronze medalists, on the other hand, compare their outcomes to those of the athletes who came in fourth place and beyond, so they tend to be more pleased with themselves than the silver medalists are, even though the silver medalists technically beat them.

From comparison to ranking and rating?

The idiom "comparing apples and oranges" refers to situations in which two items practically cannot be compared. Apples and oranges are thought to be incomparable or incommensurable. In many European languages, the phrase "comparing apples and pears" is used instead. Oranges or pears, comparison is the basis of any ranking procedure, and it has a unique role in decision making.

While we need a population of items to make a ranking based on pairwise comparisons, a score or rating can be assigned to each item individually, at least in principle. I will discuss the ranking and rating procedures.

Ranking and rating

What do we need for preparing a ranked list? First, we need a set of items (e.g., people, colleges, movies, countries, football teams); second, we need a criterion of comparison (e.g., population size, height or weight, annual income). We should be able to make clear statements for any two items A and B, such as item A is "ranked higher than," "ranked lower than," or "ranked equal to" item B. Continuing this procedure with every possible pair, a ranked list is formed. People, goods, and products have multiple features, so they can be ranked by multiple criteria. Often, different criteria are in conflict with one another: for example, price (or cost) and quality are in conflict. We cannot expect to buy a cheaper *and* more comfortable car. *Multiple-criteria decision making* thus encompasses mathematical techniques to help create ordered rankings of possible choices. For example, if a student is going to college, the decision-makers (she and her parents) have to rank the alternatives (colleges). Candidate colleges can be ranked by multiple criteria (e.g., tuition, academic status, distance from home, quality of facilities). Finally, to prepare a ranking we need an *algorithm.* An algorithm is nothing but a recipe for preparing meals—in other words, a finite list of instructions. The trick is that in order for an algorithm to work, the individual criteria should be characterized by a specific *number*, a weight, that specifies the relative importance of a criterion. Weights are subjectively determined, as I hope will become clear throughout the book. We live in a world where decision making is a combination of subjective and objective factors.[8]

By contrast, *rating* assigns a score, generally a number, to each item. In chess, for example, the Elo rating is a generally accepted system for rating and ranking chess players. Each player's strength is characterized by a number. This number is subject to change after each game—if you win against a higher-rated player it matters more than winning against a lower-rated player. We will discuss the Elo rating in chess and in other applications in the next section.

When should we use ranking and rating? A ranking question asks you to compare different items *directly* to one another (e.g., "Please rank each of the following items in order of importance, from the #1 best item through the #10 worst item"). A rating question asks you to compare different items using a common *scale* (e.g., "Please rate each of the following items on a scale of 1–10, where 1 is 'very very bad' and 10 is 'excellent' "). Both types of questions have their relevance.

Different types of rating scales exist, such as verbal (e.g., "from poor to excellent," "hate—neutral—love"), graphical (e.g., self-reporting pain graphics in a medical exam room), and numerical (e.g., grades in school, SAT scores). The Harvard psychologist S. S. Stevens in the 1940s used the words *nominal, ordinal, interval,* and *ratio* to describe a hierarchy of measurement scales.[9] Stevens claimed that all measurement in science was conducted using these four types of scales:

The *nominal* level uses just words.

The *ordinal* scale permits *rank order*, such as first, second, etc. The *relative degree of difference* cannot be seen by adopting this scale.

The *interval* type allows for the *degree of difference* between items. Temperature with the Celsius scale is a good example. Does it make any sense to say that 20° is twice as hot as 10°? Of course not.

Conventional physical quantities, such as mass, length, and duration, belong to the *ratio* type of scale. Both zero and ratio have a meaning. The duration of an event might be "twice as long" as another. (For a critique, see the notes.[10])

We are permanently faced with the problem of converting subjective qualities into objective-looking numbers.

How do you rate your pain?

Some rating scales combine these various types. When was the last time you were in a medical office? Did you see a pain scale anywhere? The Wong-Baker Faces Pain Rating Scale was originally developed to help children identify the level of their pain. It is based on a numerical pain-rating scale from zero to 10, with zero being no pain and 10 being the worst pain imaginable. The scale includes faces, written descriptions, and numbers. There are six faces in the Wong-Baker scale. The first face represents a pain score of zero and indicates no hurt at all. The second face represents a pain score of two and indicates "hurts a little bit." The third, fourth, and fifth faces represent a pain score of four, six, and eight, respectively, and they indicate "hurts a little more" and so on. The sixth face represents a pain score of 10 and indicates "hurts worst."

It is not a trivial question to ask how we should rate our pain. When we map the level of our "multidimensional" pain to an integer number between zero and nine, we compress information. Each spring, I take some physical therapy sessions for maintaining my neck mobility. It is a big help, but I always have difficulty identifying a number to characterize my own level of pain. What does it mean for pain to be zero? How should I describe the effectiveness of the therapy?

I resonate somewhat with the excellent nonfiction writer Eula Biss, who mentions five types of pain: physical, emotional, spiritual, social, and financial. A pain management website writes: "Numbers Don't Tell the Whole Story, Experts Say Better Pain Assessment Measures Needed."[11] Still, when I say to my physical therapist, Sandi, "Well, maybe three," he is able to decode my implicit message "it could have been much, much worse." Occasionally, when I have a very good day, I say, "It is really zero." What does it mean to have zero pain?

A few sentences about the history, philosophy, and cognitive science of zero

Zero was not always on the number scale. The concept of zero emerged from the contemplation of the void by the Buddhists. While the notion of emptiness has negative connotations in the realm of Western psychology, the Buddhists do not identify emptiness with the concept of nothing.[12,13,14]

The number zero was discovered (or invented) in India. ("Discovered" assumes that the concept existed independent of human activity, but we created our own label for it, while "invented" implies that zero is a human construction.) Zero appeared in the Bakhshali manuscript, denoted by a point: ·. While symbols as placeholders were used earlier by the Babylonians and Mayans, this script seems to be the first where the symbol represents "nothing" itself. Zero as a number is the result of the Buddhists' deep introspection. The Bakhshali manuscript is located in the University of

Oxford's Bodleian Libraries, and it caused a great deal of excitement among the historians of mathematics when, in 2017, radiocarbon dating of the documents showed that zero appeared in the third or fourth century, four or five hundred years earlier than it had been previously assumed.[15] Zero did not arrive in Europe until around 1200, when the Italian mathematician Fibonacci (c. 1175–1250) returned from a trip to North Africa, but now the entire digital age is based on the difference between "nothing" and "something."

Integrative studies in cognitive science that combine developmental psychology, animal cognition, and neurophysiology indicate that zero emerges in four stages: first, sensory "nothing" (i.e., the lack of any stimulus); second, the categorical "something," still qualitative; third, quantitative categorization via empty sets; and fourth, the transition from the empty set to the number zero itself.[16] Present-day cognitive neuroscience investigates the neural mechanisms of representing empty sets and zero. This is not an easy task for our brains. Neurons in our sensory systems evolved to respond to external stimuli. If no stimulus exists, the brain ought to be in a resting state. However, modern neurophysiological experiments suggest that neurons in the prefrontal cortex are able to detect actively the presence of "nothing."

Zero and non-zero, nothing and something, are basic categories in our digital age. So, we should think three times (referring to another magical number) when we declare: "My pain is zero." But as Eula Biss writes, "I'm not a mathematician. I'm sitting in a hospital trying to measure my pain on a scale from zero to ten. For this purpose, I need a zero."

As you see, ranking and rating both center about a recurring question: how objective is the ranking/rating procedure? Something is objective if it represents the external world without bias and presuppositions, while something is subjective if it results from personal preferences. Somehow, we combine the two approaches, as the next example will illustrate.

Rating graduate school applicants

During the month of December, my seasonal duty as a college professor is writing letters of recommendation and rating students based on several criteria in order to help them gain acceptance to graduate schools. Students should ask a number of professors to evaluate them. Occasionally, I have to tell a student that I would not be able to write a strong recommendation, so it would be better not to ask me. We evaluators combine quasi-objective data (say, grades) and subjective impressions to generate a rating score. Despite the subjective nature, these evaluations are far from random, and college professors don't have better ways of helping students and graduate programs find a good match. Admissions committees have a strong interest in ensuring they accept only mature, polite, reliable, and stable people into their program, and my professional duty is to help them achieve this goal.

CollegeNET is a corporation that provides software as a service to many universities, among other institutions, for admissions and applications evaluations. Their software uses six criteria to rate students:

- Knowledge in chosen field
- Motivation and perseverance toward goals
- Ability to work independently
- Ability to express thoughts in speech and writing
- Ability/potential for college teaching
- Ability to plan and conduct research

For each criterion, those rating students should choose among five options: exceptional (upper 5 percent), outstanding (next 15 percent), very good (next 15 percent), good (next 15 percent), or okay (next 50 percent). (In some other software the "exceptional" is the upper 2 percent. I have noticed that while I readily place students in the exceptional category if it is defined as the upper 5 percent,

I infrequently place students in this category if it is defined as the upper 2 percent.)

How do we generate the numbers and choose the appropriate rubric? In principle, a micro-rationalist, bottom-up approach would work: teachers could collect and store data from students throughout decades, and they might have a formal algorithm for calculating the percentages. I believe it is more likely that many of us apply top-down strategies. I ask myself: Do I want to grant a set of grades that is all "exceptional"? Does the applicant have a clear weak point, in which I should check the third or maybe the fourth category? What if I score four outstanding and two exceptional? Good or bad, decision-makers calculate the sum of the grades, analyze the grade distribution, and then make some subjective analysis of how to make recommendations. As Churchill might have said: Quantification is the worst form of evaluation, except for all the others.

I describe now two celebrated examples in which quantification works well: the rating and ranking of mathematicians and chess players are well accepted by their communities.

From the ranking of mathematicians to the rating of chess players

Erdős number

Saffron is the best Indian restaurant in Kalamazoo, Michigan, and my wife and I dined there recently this spring with close friends. (If you are from Budapest and live in southwest Michigan, the best candidate for a close American-born friend is someone with a wife who is, if not from Budapest, then from Prague. My friend Tom was born and raised in Detroit—much before the riot in 1967—and has an excellent sense about the Central European spirit.) As I entered, I asked Tom: "I see a guy here that should be from your math department—he has a car with a vanity plate that reads 'Erdős #1.' Do you know him?" The car happens to be owned by Allen Schwenk, who is among the 512 mathematicians who co-authored with Paul Erdős (1913–1996). Schwenk and Erdős co-authored four papers in the field of graph theory, a subfield of math that was so popular at Western Michigan University in Kalamazoo that Erdős traveled there frequently. Thirty years after their collaboration, Schwenk still speaks with great enthusiasm about Erdős's influence, which one can imagine after seeing his license plate.

Erdős published around 1,500 mathematical articles during his lifetime,(actually the last paper was published in 2015, almost two decades after his death), most of which were co-written. He had 512 direct collaborators; these are the people with an Erdős number of 1. The people who have collaborated with them (but not with Erdős himself) have an Erdős number of 2 (around 10,000 people), and those who have collaborated with people who have an Erdős number of 2 (but not with anyone with an Erdős number of 1) have an Erdős number of 3 (people such as myself).

In the context of ranking, it is not only a nice story about how mathematicians accept, more than semi-seriously, the importance of the Erdős number, as a measure of a mathematician's nobility, but it also gives a unique example for the self-organizing mechanism of a wise and democratic community of the mathematicians.

Bridges to connect mathematicians to neurobiologists, economists, and even philosophers?

János (John) Szentágothai (1912–1994) (JSz), one of the most distinguished neuroanatomists of the 20th century, has an Erdős number 2, since he co-authored a paper with Alfréd Rényi (1921–1970), published in 1956 (about the probability of synaptic transmission in Clarke columns). It seems to be a plausible hypothesis that JSz is a bridge to connect the community of mathematicians to neurobiologists and even to philosophers. JSz co-wrote a book with two other scientific nobilities, Nobel Prize winner Sir John Eccles (1903–1997) and neurophysiologist Masao Ito (1928–2018). (It is interesting to note that JSz himself was thinking about the graph of the network of the cerebral cortex in terms of what is today called a "small world." JSz hinted that the organization of the cortical network should be some intermediate between random and regular structures. He estimated that any neuron of the neocortex is connected with any other by a chain of not more than five neurons on average.) Eccles has co-written a book with Sir Karl Popper (1902–1994),[17] so there is a direct math-neurobiology-philosophy chain.

Another non-mathematician with Erdős number 2 via Rényi is András Bródy (1924–2010), a Hungarian economist. They also published a paper in the same memorable year of 1956 (this one was about the problem of regulation of prices) (Figure 2.1). So, another question is suggested: Since most likely all people with Erdős number 1 are mathematicians, how many non-mathematicians have Erdős number 2, and how many other scientific communities are involved in the collaboration graph?

Rating chess players: a success story

Arpad Elo (1903–1992) was born in Hungary and moved with his parents to the United States when he was 10. He was a physics professor at Marquette University in Milwaukee, Wisconsin, and he was also the founder of the US Chess Federation. Elo created a rating system to characterize the relative strength of chess players, which was mentioned earlier in this chapter. The stronger chess player generally beats the weaker one, but not always. A chess game in the Elo rating system has an *expected score*, and the larger the difference between the two players, the smaller the chance of the weaker player's success. If a strong player loses against a player

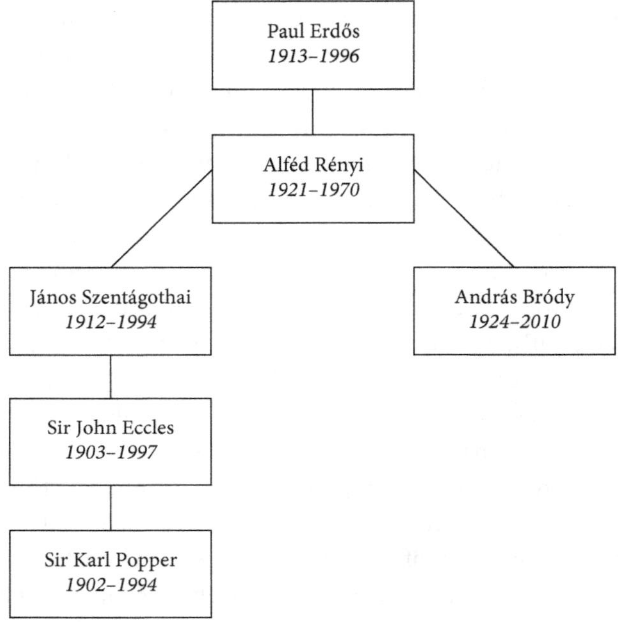

Fig. 2.1 This graph shows how Rényi's relationship with JSz acts as a bridge between the community of mathematicians and neuroscientists and how Rényi's relationship with András Bródy connects mathematicians and economists. Were the author of this book to be included in this diagram, he would be represented as a node connected to JSz.

with a very low score, the strong player's points will be significantly reduced. To avoid math, I used the very qualitative terms "strong" and "weak," but the Elo system precisely defines these terms.

The Elo system was adopted by the World Chess Federation (FIDE) in 1970 and became quite popular after its introduction. The rating system, as it is now specified, implies that a 100-point difference predicts a 64 percent change of the higher-rated player winning and a 36 percent chance of the lower-rated player winning. When I checked on September 6, 2017, the Norwegian world champion Magnus Carlsen led the list with 2,827 points, and

Vassily Ivanchuk, the Ukrainian chess grandmaster and former World Rapid Chess Champion, was ranked #32 with 2,727 points.

In 2015, there were about ten thousand players in the world whose Elo rating exceeded 2,200. This number corresponds to the level "Candidate Master." It is fair to say that the professional level of game play starts here. FIDE updates its ratings list monthly, so the list of Candidate Masters changes relatively frequently.

The Elo rating system has been adopted outside of chess to rank players in various games (from Scrabble to backgammon to Go to baseball to rugby to online games). You also will find an example for a nontrivial application (i.e., to measure social dominance) in Chapter 3.

The Elo rating is a well-functioning system, but of course it can be subject to improvement. Mark Glickman, a mathematician from Harvard, has suggested a method that takes into account the *reliability* of one's rating. One's rating would not be considered reliable if one had not played for a long period of time. The Glicko system, as he called it, extends the Elo system by computing not only a rating but also a "ratings deviation" (RD), which measures the uncertainty in a rating (high RDs correspond to unreliable ratings).

From the Ten Commandments to the top-10 mania

Lists

Our love of ordered lists is in its heyday, but it may be older than it seems! Let's go back to the distant past. The Ten Commandments appear at first to be an unranked *list*. However, in the rabbinic literature there are different interpretations regarding whether some items are more important than others. For example, Rabbi Yehudah ha-Nassi[18] has said: "Be as scrupulous in observing a minor commandment as a major commandment, because you do not know the value of each commandment" (Pirkei Avot 2: 1) (Pirkei Avot is generally translated as Ethics of the Fathers). However, for others, the situation is quite complicated. Rabbinical Judaism refers to the 613 commandments (mitzvah) given in the Torah at Mount Sinai and the seven rabbinic commandments instituted later, for a total of 620 commandments. According to some, "There IS a value to each mitzvah; we just don't know what it is. A specific mitzvah may be worth dozens of other mitzvot. Only the Master of Opinions knows how the comparison between sins and merits is made" (Rambam, Mishneh Torah, Hilchot Teshuvah 3: 1–2). Our obsession with charts, rankings, top-10 lists, etc. might be considered some secular echo of the litanies of faith.

Much later in history, Martin Luther (1483–1546) wrote the 95 Theses, published as a poster on the door of the Schlosskirche (Castle Church) in Wittenberg, Germany, on October 31, 1517, which initiated the Protestant Reformation. The abuse of indulgences (a way to reduce the amount of punishment one has to undergo for sins) had become a serious problem that the Catholic Church recognized but was unable to handle and caused the greatest crisis in the history of the Western Christian church. The original Latin version of Luther's 95 Theses was soon translated into German and was printed and circulated. We cannot underestimate the significance of the availability of the printing press in reproducing and propagating the text. "The medium is the message," as Marshall McLuhan (1911–1980) famously proclaimed, and the 95 Theses quickly became the symbol of protest.

In addition to alluding to the role of lists in the history of religion, it is useful to explain our inherent love of ordered lists. BBC E-cyclopedia defines listmania as "media obsession to categorize anything into lists, be they musical artists, memorable sporting moments, quotations, words of the year, etc." Our brains and mind love lists, and the ubiquity of lists is evident everywhere online. I spent six minutes at cnn.com and generated a list of lists accessible with at most one click:

- Eight best Istanbul hotels
- Five ways you're losing money without even realizing it
- Seven best places to stay in Napa Valley
- 12 amazing hotels perfect for animal lovers
- The best photos of the solar eclipse
- 10 of the best beaches near airports
- Eight tips for surviving long flights
- Four questions to ask yourself before retiring

Lists in a brain game

Our brain's function is to process external information perceived by all of our sensory systems. The incoming information is useful only if we are able to comprehend it, and lists help us organize information both new and old. There are situations when people are in complex, dynamic environments that demand they rapidly understand what is happening so they can make decisions followed by actions. Historically, military command and control is a field from which the theory and practice of *situational awareness* has emerged. However, other activities, such as air traffic control, firefighting, or aviation, and more ordinary complex tasks, like driving a car or even riding a bicycle, require us to comprehend rapidly changing environments and react in a timely fashion. Situational awareness starts with the *perception* of environmental elements and events with respect to time or space, followed by the *comprehension* of their meaning and by the *projection* of possible future events.

Lists help us comprehend incoming information. *Kim's Game* is a famous example of how a complex environment should be mapped into a list, and how to improve the efficiency of the comprehension. Bert and Kate McKay, founders of the *Art of Manliness*,[19] summarized the origin of the game so nicely that I quote it here:

In Rudyard Kipling's famous novel *Kim*, Kimball O'Hara, an Irish teenager, undergoes training to be a spy for the British Secret Service. As part of this training, he is mentored by Lurgan Sahib, an ostensible owner of a jewelry store in British India, who is really doing espionage work against the Russians.

Lurgan invites both his boy servant and Kim to play the "Jewel Game." The shopkeeper lays 15 jewels out on a tray, has the two young men look at them for a minute, and then covers the stones with a newspaper. The servant, who has practiced the game many times before, is easily able to name and exactly describe all the jewels under the paper, and can even accurately guess the weight of each stone. Kim, however, struggles with his recall and cannot transcribe a complete list of what lies under the paper.

Kim protests that the servant is more familiar with jewels than he is, and asks for a rematch. This time the tray is lined with odds and ends from the shop and kitchen. But the servant's memory easily beats Kim's once again, and he even wins a match in which he only feels the objects while blindfolded before they are covered up.

Both humbled and intrigued, Kim wishes to know how the boy has become such a master of the game. Lurgan answers: "By doing it many times over till it is done perfectly—for it is worth doing."

Over the next 10 days, Kim and the servant practice over and over together, using all different kinds of objects—jewels, daggers, photographs, and more. Soon, Kim's powers of observation come to rival his mentor's.

Today this game is known as "Kim's Game" and it is played both by Boy Scouts and by military snipers to increase their ability to notice and remember details. It's an easy game to execute: have someone place a bunch of different objects on a table (24 is a good number), study them for a minute, and then cover them with a cloth. Now write down as many of the objects as you can remember. You should be able to recall at least 16 or more.

Remembering lists

The human brain generally does not have the ability to remember long lists of unstructured items. We aren't very good at remembering a series of numbers, of nonsense words, or of goods to purchase in the supermarket. One of the pioneers of memory research, Hermann Ebbinghaus (1850–1909), made memory studies around 1885 on himself and tried to memorize nonsense syllables. Time and again, he tested his memory and realized that the quality of his memories decayed exponentially, and he theorized that the performance of his memory was quantitatively characterized by what is called the "forgetting curve." He also found that his performance depended on the number of items, and it was more difficult to memorize long lists of items as opposed to short lists.

There are big exceptions to these generalities. Some people are able to remember lists of nonsense items for literally decades. Alexander Luria (1902–1977), a Soviet neuropsychologist, studied a journalist named Solomon Shereshevski (1886–1958), who apparently had a basically infinite memory. He was able to memorize long lists, mathematical formulae, speeches, and poems, even in foreign languages, and recall these lists 14 years later just as well as he had on the day he learned them. His performance did not depend on the length of the items, deviating from the theory suggested by Ebbinghaus's observations. Shereshevski was diagnosed with synesthesia, a neurological condition in which different senses are coupled. When he realized his ability, he performed as a mnemonist. Despite the allure of having a perfect memory, his abilities also created disorders in his everyday life, as it was difficult to him to discriminate between events that happened minutes or years ago.[20] Luria had a strong influence on the famous neurologist and writer Oliver Sacks (1933–2015).

Everyday people use *mnemonic* techniques to improve their memory performance. One technique, called *spaced repetition*, is still popular and goes back to Ebbinghaus's research: "with any considerable number of repetitions a suitable distribution of them over a space of time is decidedly more advantageous than the massing of them at a single time." (This is why we tell our students not to cram!) Additionally, later in the learning period, forgetting happens more slowly, so you can slow down the pace of the rehearsal. Spaced repetition has proven to be one of the most efficient ways of learning items for long-term retention.[21] If you were asked to try to recall the name of the person shortly after you had just met her and then again after a longer interval, it would be a good strategy to space out the repetition and recall the name after five minutes, 30 minutes, and then after two hours, instead of trying to recall the name every 30 minutes.

A number of experiments have studied efficient methods of learning in medical schools, where an excessive amount of factual knowledge (frequently in the form of lists) is required learning. Trials in expanding study intervals on days 1, 6, 16, and 29 were significantly more efficient that studying in a steady interval (days 1, 10, 20, 29) in specific cases.[22] But, students, first you should understand concepts before memorizing them!

My mathematician friend from Budapest (John, as you may remember, who never had a car) suggested that I write about a type of software used by one of his sons, who serves as an interpreter for the European Union and likes to learn new words in many languages. The program, called Anki, adopts the old method of physical flashcards.[23] I don't know much better than (cautiously) to trust in the wisdom of the crowds, as I will discuss later in this chapter (but this does not necessarily mean that I am happy with the results of some elections in the last several years). So I checked Reddit, "the front page of the Internet," as the site bills itself, and found an inquiry:

> 1. How does it fare over a long period of time? 2. Do you truly retain (mostly) everything you've attempted to retain over the

> course of your use of it? 3. How would you rank the program from 1–10 on a scale of how much it's impacted your studies?

As you see, one question is about ranking (of course, to request rating or scoring would have been more precise), and here is an answer:

> Purely anecdotal, but . . . 1) After about 5 months I have learned many more Russian vocabulary words than I would have otherwise; 2) So far, yes; 3) 8–9; I feel more confident than ever that I have in place an effective tool to review my material and not let any of it slip through the cracks. Lord, how I wish I had discovered Anki before law school.

Love of lists

Claudia Hammond, a British psychologist turned BBC broadcaster, wrote a piece titled "Nine psychological reasons why we love lists."[24] The title itself is a provocation: Are there really nine, and not seven or 13, reasons? Dear reader, do you have your own list of reasons we love lists? In any case, Hammond's list is as follows:

1. We know exactly what we're getting.
2. We don't like missing out.
3. They feel less taxing on the brain.
4. We like to think we are too busy to read anything else.
5. They are easy to scan for information.
6. We always know how much is left.
7. It's fun to try to guess what's on the list.
8. We love being proved right.
9. A list feels definitive.

Most likely, this is not an ordered list. However, if I see a list titled "25 best liberal arts colleges," I know for sure that this list will contain 25 items. Some lists use reverse order, where the winner comes last. Ten is a nice number, and the top 10 lists on David Letterman's show were a popular element of late-night television for a while. In one piece, "Top ten numbers between one and ten," with the contribution of Casey Kasem (co-founder of *American Top 40*), Letterman parodied the nonsense of ranking. (If you don't know, or even if you do, see the video in the notes.[25])

Umberto Eco (1932–2016), the celebrated Italian novelist and a public intellectual, famously wrote: "We like lists because we don't want to die," and lists are means of grasping the incomprehensible. Whenever we encounter new information, we subconsciously generate lists to organize it. Eco found lists to be important as a way of escaping thoughts about death. My profane observation is that while we are preparing to-do lists, we are alive.

To-do lists

Many of us prepare "to-do" lists—prioritized lists of all the tasks that we need to carry out generally "soon." So first we make a list of everything that we have to do, and then we make a ranked list with the most important tasks at the top of the list and the least important tasks at the bottom. It is not as simple as it sounds to prepare a to-do list, and we might ask whether we have some "best" algorithm of constructing one. There are different features of tasks we have to do: urgency, expected penalty for postponing, time needed to complete the task, etc. You certainly cannot postpone picking up your kid from kindergarten. And if your boss asks you to give your quick opinion about a situation (maybe in the form of a list) at noon, you will have to decide whether you do it before or after lunch (well, an eager beaver could do it *instead* of lunch). Some people believe that having a long to-do list is proof of their value and indispensability. (Of course, it is a sad fact that cemeteries are full of indispensable people, and successful people are often able to outsource their tasks, as most famously Tom Sawyer did with the whitewashing of the fence.)

It is reasonable to have to-do lists for different time scales, for short-term, intermediate-term, and long-term projects. "Short-term" might be one day, or, in busy periods, maybe two hours. We should write down things, and it is useful to use pen and paper to do so (used envelopes are very good for this purpose!). Our conscious mind is able to keep no more than four or five things in attention at once, and generally we have more than four or five things to do during the day. (Can you write down how many things you have to do today, or, if you read this paragraph in the late evening, then tomorrow?)

A hundred years ago, Ivy Lee, a business consultant, was asked to improve the efficiency of steel magnate Charles M. Schwab's business. Lee asked for 15 minutes of discussion time with each executive.

"How much will it cost me?" Schwab asked.

"Nothing," Lee said, "Unless it works. After three months, you can send me a check for whatever you feel it's worth to you."

He suggested a seemingly simple technique to each executive:[26]

1. At the end of each workday, write down the six most important things you need to accomplish tomorrow. Do not write down more than six tasks.
2. Prioritize those six items in order of their true importance.
3. When you arrive tomorrow, concentrate only on the first task. Work until the first task is finished before moving on to the second task.
4. Approach the rest of your list in the same fashion. At the end of the day, move any unfinished items to a new list of six tasks for the following day.
5. Repeat this process every working day.

The technique worked, and Lee got a check for $25,000. Multiply this number by 15 to calculate its equivalent today. Since I am very modest, send me only the original amount if the technique works for you.

Warren Buffet has his own trick for managing priorities. In a famous story, he asked his personal airplane pilot, Mike Flint, to list 25 things he wanted to do on List A. Then he made him circle the top five among these to make List B. Here's how the conversation went:[27]

FLINT: "Well, the top five are my primary focus, but the other twenty come in a close second. They are still important so I'll work on those intermittently as I see fit. They are not as urgent, but I still plan to give them a dedicated effort."

BUFFETT: "No. You've got it wrong, Mike. Everything you didn't circle just became your Avoid-At-All-Cost list. No matter what, these things get no attention from you until you've succeeded with your top five."

So, prioritization and attention allocation are the main elements of both individual and institutional decision making. (For the attention allocation of political systems, see the book[28]). While to-do lists have proven to be useful for helping us organize our activities, the structure provided by lists has become a popular form in the written media, and this genre is called the *listicle*.

Might-do list: A new silver bullet?

While to-do lists are now generally accepted as a valuable strategy for time management, they also have a less favorable side: they reward completing small tasks. You can spend your time managing small tasks, assigning 30 seconds to 60 minutes to execute a task, but do we do anything big and worthwhile? John Zeratsky advertises himself as "helping people make time for what matters."

He uses the concept of "One Big Thing" to make a day more productive and to plan not only his days but also his weeks and months. He starts by designing his daily activity from a "might-do" list, and he suggests a two-stage strategy: first, from the might-do list choose One Big Thing of the day, and second, design the allocation of personal resources (time and energy) to the One Big Thing and to the other obligations. The advantage of this two-stage strategy is that it separates the *planning* and the *doing* processes. When you are tired during the doing process, don't change your plan! You should trust that the planning process was reliable![29]

From Al Capone to the listicles

Al Capone (1899–1947), the infamous boss of an efficient organized-crime empire, was officially called "public enemy number one" in 1930 by the Chicago authorities. As we know, "there is nothing new under the sun," and even in the Roman times, Cicero (106–43 BCE) used the notion of public enemy—*hostis publicus*. The Chicago Crime Commission released a list in 1930 of 28 men labeled "public enemies," and Capone's name was on the top of the list. He also leads the list "The 17 most notorious mobsters from Chicago,"[30] as he managed to combine the characteristics of a mobster with the fame of a pop star. It is not surprising that history.com published a listicle with the title "8 things you should know about Al Capone." Totally accidentally, an article in the magazine of the University of Chicago (written by the linguist Arika Okrent) nicely explains that a listicle is a literary form, similar to a limerick or haiku. If you see a number in the title of the listicle, you already know an important bit of information about the quantity you are supposed to receive. You could decide, "Yes, I am ready to spend a specified, affordable amount of time to know the contents of this list." Probably still the number 10 appears most frequently in the titles, but other numbers are often selected to make the genre a little more fun. Listicles provide ordered lists, so if the title announces "the best of," "the most of," or "the worst of," we know that something or somebody will be listed. Our brains like the flow of linearly arranged items, so we buy it.

Writing about listicles and haiku, I can't resist publishing my first haiku:

Three lines—one listicle
Our brains like lists
the number of items known
oh, the end is here.

From the cognitive biases of the individuals to the wisdom of crowds and back

As I am thinking about my own cognition when I am assigning scores to students' knowledge, motivation, and ability, there would be no sense in denying the subjective elements of my evaluation method. Somehow, I integrate my memories about the student's character, attitude, and performance. Of course, with close students I have had numerous conversations about very different aspects of life, from work ethics to philosophy of science, and from politics to love. I try to be objective, but it is difficult to avoid what is called the *halo effect*. The halo effect is a form of cognitive bias in which our overall impression of a person determines our evaluation of specific traits and performance. The emergence of the concept goes back to Edward Thorndike (1874–1949), a psychologist who described the concept in a study published about a hundred years ago as relating to the way that commanding officers rated their soldiers. Since I became aware of the halo effect, I make more effort to rate each item independently from all other items. Fortunately, a student is evaluated by several other people, so maybe (yes, maybe) the individual biases average out. Collective wisdom is supposed to be more efficient than individual judgment, as I will discuss now.

Francis Galton (1822–1911), a half-cousin of Charles Darwin, loved to count and measure everything. While he has a bad reputation for introducing the field of eugenics with the goal of supposedly improving the genetic quality of the human population, he contributed to making the fields of biology, psychology, and sociology more quantitative. A famous story reports that he visited the West of England Fat Stock and Poultry Exhibition, where, among other animals, an ox was on display. He asked the guests to estimate the weight of the animal. About eight hundred people participated, and the *median* estimate was very, very close to the real value. (The median value is the value that separates the higher half from the lower half of a data sample.) The take-home message of this observation is that the accuracy of the estimate of a population exceeds the accuracy of the estimates of individual experts. This notion is called and popularized as *the wisdom of the crowd*, which was the title of a book by James Surowiecki in 2005.[31] We don't have to believe that the opinion of the crowd is impeccable. Surowiecki argued that the estimation of the crowd is really good if individual opinions are independent. Independence, however, seems to be an illusion. Nietzsche recognized and sharply criticized the herd instinct we humans have. If we let ourselves be influenced by others (led by others like a sheep, as Nietzsche writes), then the crowd's calculation leads to biased results. This has been demonstrated by the works of a leading computational social science group in Zurich, Switzerland, directed by Dirk Helbing. They gave several neutral questions to people, who had to estimate some data related to demography or crimes (e.g., population density, number of rapes in a given year in Switzerland). If the participants did not communicate with each other, they got a better result than when they could exchange opinions with each other. In fact, when opinions were shared, the range of estimates was reduced and the center of opinions shifted away from the real value. Their finding was surprising.[32] Generally we believe that consensus implies better decision making; however, it might happen that initially small deviations from the "good" value are amplified by the herd mechanism.

What we see is that if opinions are distributed over a larger range, the estimate is better. Along the same line, a diverse population of problem-solvers creates better decisions than a group of more uniform, well-performing solvers, as the model calculations of the complex systems scientist Scott Page from the University Michigan have demonstrated.[33]

If a crowd is clever, how big should be the size of committees? Naïve intuition might suggest the larger, the better. However, many of us will have second thoughts if we remember the saying "A camel is a horse designed by a committee." Social dynamics and complex systems scientists have studied real-world situations[34] and tried to determine things like the number of (1) political experts a journalist should consult to predict election outcomes; (2) doctors a patient should consult to obtain optimal accuracy of a diagnosis; and (3) economists a government should ask to make a good estimate about the future trajectory of the economy. Both mathematical analyses and experimental data suggest that smaller crowds outperform larger crowds, and groups with moderate size (five to 15) produce better results than large committees.

Let's take a step back. Can we consider that even an individual might be a crowd? First, there are people who might have more than one opinion about something or somebody. Also, people may give different estimates several weeks later. As it turned out that averaging is useful, even individuals may benefit by integrating their different perspectives: crowds and crowdsourcing may exist within a single mind!

Lessons learned: the basic concepts

Comparing ourselves to others is an elementary human activity, and we cannot avoid making comparisons and being compared.

There is a tradeoff: favorable comparisons make us happier (at least in the short term), but unfavorable ones drive us to make things harder. Systematic comparison among many elements gives a ranked list. Rating is, in principle, simpler—a score (generally, but not necessarily, a number) is assigned to the object or subject being rated, independently of the scores assigned to other objects or subjects. As a teacher who gives grades, of course I know that it is simply impossible to always be objective: there is some interaction among the grades of individual students. Ordered lists are based on the rankings of elements. Somehow we love, read, and prepare lists, since they condense and organize information. Like it or not, each day we read a good number of ranked lists, many times in the form of a listicle, a style that bloggers and journalists recently adopted to convey information via the ranking procedure.

Now we are ready to discuss the biological and social mechanisms of the ranking processes and even the computational algorithms associated with these mechanisms. Specifically, the next chapter is about social rankings occurring in both animal and human societies.

3
Social ranking in animal and human societies

Pecking order in chicken community

You can discover scientific facts when you are as young as 10 years old, as the incredible story of the discovery of pecking order among chickens suggests. Thorleif Schjelderup-Ebbe (1894–1976) grew up in a flourishing family in Oslo, Norway. The family spent the summers in a suburb, where their house had a yard full of chickens, which became a source of great interest to young Thorleif, especially their social relationships. According to a family story, he made observations and notes regarding the manner in which chicken A masters chicken B, and chicken B masters C, and so on, and he coined the term "pecking order" to describe the hierarchy he observed among chickens. Chickens not only rank themselves as a community, but they also accept their places in the ranking. The hierarchical order prescribes the priority of access to resources, especially food and mates. If you are a random chicken, neither a "top chicken" nor a "bottom chicken," you will accept that the top guy comes first, and will avoid superfluous conflict. And when a lesser chicken is around your selected mate, he will know not to overstep.

Despite his brilliant discovery about social hierarchy among chickens, Schjelderup-Ebbe was not very successful at navigating human hierarchy. As a college student, he was heavily influenced by the first female professor in Norway, Kristine Bonnevie. However, since she erroneously believed that an article criticizing her was written by Schjelderup-Ebbe, she withdrew her support from his research, and he never managed to obtain a favorable reputation in his own country again. Still, Schjelderup-Ebbe's concept of pecking order led to detailed studies of dominance hierarchy in a variety of species, ranging from insects to primates. What works for chicken society somehow works too for humans, even though we have a more complex social organization. How?

Measuring dominance and understanding the formation of hierarchies

Observing animal behavior

There is a long tradition of observing and recording animal behavior. The earliest example comes from cave paintings, and according to archaeologists, the oldest cave painting identified so far resides in India and is at least 35,400 years old. It depicts a pig; in fact, the most common subjects of cave paintings are large wild animals. In written history, the *History of Animals* by Aristotle (384–322 BCE) contains many accurate eyewitness observations. However, at that time, the continuous observation of the social behavior of animal groups in their natural environments with the least possible intervention proved to be very difficult. Contemporary ecologists and ethologists use wireless sensors and global positioning systems (GPS) to track and monitor the behavior and interaction of freely moving animals.

The emergence of dominance hierarchy: self-organization

Linear dominance hierarchies proved to be very efficient for community resource management in a wide variety of social animals, from insects to fish and from birds to primates. Since more and more data have been accumulated, it has become possible to test hypotheses in contemporary animal behavior studies about the mechanisms behind the formation of evolutionary hierarchies. One famous enterprise is the Amboseli Baboon Research Project,[1] which obtains and analyzes data on the behavior of wild baboons. Tens of thousands of observations of agonistic encounters have been made. The repeated encounters have winners and losers, so basically the individuals participate in a tournament. Surprisingly, animal behavior researchers have used the Elo rating method to analyze the results of past "games" and predict the outcome of the future ones. The so-called winner and loser effect seems to be convincing. It describes the phenomenon in which winners tend to become more likely to win in subsequent encounters and losers tend to become more likely to lose.[2,3]

Behavioral studies on parakeets have also led to a new hypothesis concerning knowledge of social rank and its relationship to aggressive behavior. Parakeets have some features, such as large brain size and relatively long lifespan, that make them appropriate subjects for studying complex social behavior. When a group of parakeets has just been formed, the social group exhibits no structural behavior. After a week or so, their behavioral strategies begin to change. First, the animals learn their ranks after both observing and participating in a number of aggressive interactions, meaning they form some social memory. Second, they use their knowledge of their rank in their decisions and in subsequent actions. These birds make decisions regarding whom they should fight with (or against) and with whom they should not, based on their knowledge of social hierarchies. Parakeets not only avoid fights with those ranked higher than themselves, but they also don't waste their energy fighting with those ranked significantly lower than themselves.[4]

Dominance hierarchies certainly limit the escalation of conflicts and contribute to the maintenance of social stability.

Two ways to the top: brute force versus knowledge

Evolutionary mechanisms: dominance and prestige

What did our ancestors need for survival? The same as we do. Food and mates! Evolutionary mechanisms led to the formation of hierarchies to regulate access to these resources.

Some biological mechanisms that determine the rank of an individual among peers are similar in both primates and humans. Individuals at the top of the hierarchy benefit from their higher social rank, as they subsequently have more resources with which to cultivate a healthier and happier life. The desire to achieve a higher social rank appears to be a universal driving force for all human beings.

There are two distinct mechanisms for navigating the social ladder, dominance and prestige. Dominance is an evolutionarily more ancient strategy and is based on the ability to intimidate other members in the group by *physical size* and *strength*. In dominance hierarchies, the group members don't accept the social rank freely, only by coercion. Members of a colony fight, and the winners of these fights will be accepted as "superiors" and the losers as "subordinates." The hierarchy formed naturally serves as a way of preventing superfluous fighting and injuries within a colony.

Prestige, as a strategy, is evolutionarily younger and is based on *skills* and *knowledge* as appraised by the community. Prestige hierarchies are maintained by the consent of the community, without pressure being applied by particular members. It is not a surprise that those with different personality traits adopt different strategies. People using dominance to secure their status tend to be more aggressive, manipulative, and narcissistic. By contrast, people who use prestige instead tend to be more conscientious, confident, and diplomatic. Both strategies might have some negative consequences. Dominant leaders place a higher priority on maintaining power than achieving group goals, while leaders with prestige sometimes prioritize their social approval over group goals.[5]

The topic of dominance versus prestige is employed frequently in movies and books. Usually negative characters use dominance, and positive participants develop prestige, to achieve their ends—for example, Darth Vader and Master Yoda in *Star Wars*, or Scar and Simba (or Mufasa) in *The Lion King*. In the latter pair, Scar ruled by domination over other animals, as he had support from numerous hyenas, much like a military, while Simba (or Mufasa) was supported based on his prestige and the respect he earned in his community.

I have dual citizenship in the United States and Hungary, and the leaders of my two countries are currently (2018) the champions of populist authoritarian leadership. They are aggressive and narcissistic and possess questionable moral character. Social psychology suggests that under psychological threat, there is an increase in the appeal of an external agent (from God to the president) who could help individuals cope with the threat. Even when there is no real threat, external agents can artificially generate the feeling that a group is being threatened, hence the strategy of stoking the fear that "we must not allow in a single migrant!" They build fences and walls and then suggest Brussels and Mexico must pay for the border costs. This strategy is particularly productive when people feel uncertainty and have the psychological sense of lacking control over their own lives.[6,7]

The biological machinery behind social ranking

From sociobiology to evolutionary psychology

Edward Wilson, the famous biologist and writer, explained altruism, aggression, and other social behaviors in terms of biological evolution. His 1975 book on what he called *sociobiology*[8] dealt mostly with social animals (such as ants), and it contained a single chapter on humans, which provoked sharp debates. The opponents of sociobiology were headed by leading evolutionary biologists Richard Lewontin and Stephen Jay Gould (1941–2002), who attacked sociobiology for supporting biological determinism. Biological determinism may have, as they argued, serious negative social consequences. Sociobiology has been replaced by *evolutionary psychology*, a less direct, more neutral theory for explaining the evolution of human behavior and culture by mechanisms of natural selection.[9] There is no reason to deny our biological roots, and we will discuss their role in forming hierarchies.

Hormones, stress, and ranking

Levels of the hormone testosterone are a good measure of social dominance both in monkeys and humans. Higher testosterone levels have been measured in socially dominant individuals as compared to those of socially inferior individuals in experimental studies. Experiments have also suggested that victory (or defeat) implies an increase (or decrease) of testosterone levels in male athletes (observed not only in football, rugby, tennis, and wrestling, but also in chess). Changes in testosterone levels have been measured not only in athletes but also in fans of sporting events.

Three hormones—adrenaline, cortisol, and norepinephrine—show correlation with stress. Is it good to have some stress? Partly, yes. Stress helps us animals survive in a dynamic environment. As new threats arise, an animal must be able to quickly perceive, comprehend, and assess the situation before making plans and acting accordingly. When an animal is stressed, its pituitary gland and adrenal cortex release stress hormones. These stress hormones will then have a number of physiological effects on the animal's body, such as an increased heart rate, increased muscle tension, and suppressed digestion and reproduction. Naïvely, it was once believed that the subordinate members of a group were the ones to display the greatest levels of stress hormones. This was imagined to be attributable to the stress of losing a fight or not having access to priority resources. Data on both monkeys and humans, however, are at best controversial. Highly ranked animals also show high levels of stress, but its duration is short-lived and helps them win the next competition (which implies further increase in their rank). Lower-ranked males, who are subject to bullying, have a chronically elevated level of stress hormones, which is really harmful and leads to further reduction of their social rank. We all know that it is very difficult to break such kinds of vicious circles.

Here is a listicle, "Eight reasons a little adrenaline can be a very good thing":[10]

1. It might help you on a deadline.
2. Your vision gets better.
3. You'll breathe more easily.
4. Other experiences are heightened.
5. It can block pain.
6. It can boost your immune system.
7. You'll get to tap into a little extra strength.
8. It might help slow aging.

We already know that lists are just lists. I don't read this listicle as the "final truth," but as an educated opinion.

Hunting skills resulted in reproduction success

Anthropological research has documented the relationship between male hunting skills and reproductive success in various societies around the globe, including the Aché, a group of hunter-gatherers living in eastern Paraguay; the Hadza, an indigenous ethnic group from northern Tanzania; the !Kung, hunter-gatherers indigenous to the Kalahari Desert in Angola, Botswana, and Namibia; the Agta, an island people indigenous to the Philippines; and the BaYaka, a nomadic people inhabiting the Central African Republic and Congo.

Eric Alden Smith, an anthropologist from University of Washington, analyzed the possible causal mechanisms. He found, among others, that the hypothesis based on the naïve expectation that "better hunters have better-fed wives and children, thereby enhancing spousal fertility and/or offspring survivorship" is not necessarily justified. In fact, the data seem to indicate that hunting is used as a status symbol rather than a main source of food.[11]

Improving versus maintaining status

There is a tradeoff between our intentions to improve our status and to maintain a stable position in the status hierarchy. A promotion means you climb to the next rung on the ladder, while getting fired sets you back one or many rungs. The overwhelming access to information in our everyday lives makes our social identity readily available to others. Actions become more easily public, so people can gain and lose social standing more easily. It is enough to think of the rapid loss of status experienced by cultural and media icons as women have bravely come forward to assert "me too" in regard to sexual harassment and assault. I am not sure where this whole movement will lead. Turning women and men into hostile, opposing sides of a battlefield is not going to be beneficial for either sex. There is something to be said, however, for the importance of revealing instances of sexual harassment and discrimination wherever it occurs. If an ideal mixture of competition and cooperation is as close to a silver bullet as we might get, as I learned from a close female colleague, creating safer environments for women may better enable cooperation and ensure that competition occurs on a level playing field. I have to admit, though, I am not totally sure how such a safer environment should look.

Martin Nowak and Karl Sigmund[12] have offered a mathematical model to show that cooperation can emerge even if recipients have no chance to return assistance to their helper. This is because providing assistance improves reputation, which in turn makes one more likely to be helped. This *indirect reciprocity* can be modeled as an asymmetrical interaction between two randomly chosen players. The interaction is asymmetrical, since one of them is the "donor," who can decide whether or not to cooperate, and the other is a passive recipient. However, the result of the decision is not localized; it is observed by a subset of the population, who might propagate the information. Consequently, the decision to cooperate might increase one's reputation, and consequently, those people who are considered more helpful might have a better chance of receiving help. The calculation of indirect reciprocity is certainly not easy. An individual considering cooperation is more likely to pursue this strategy with another cooperative individual than with an individual who will not reciprocate. The probability of knowing someone's reputation should be larger than the cost/benefit ratio of the altruistic act. Evolutionary game theory suggests that indirect reciprocity might be a mechanism for evolution of social norms. We will return to this topic in Chapter 7 about reputation.

Physical size linked to social status

If you want to be a colonel or even a president, be tall! Maybe it is not exactly true, but the "status–size hypothesis" suggests that there is a positive correlation between physical size and social status. How important is the physical size in our life? Generally we don't beat up our rivals (that is, if we don't box, as Muhammed Ali did), certainly not literally (not even then, when we are the stronger). Maybe in violent gangs, physical confrontation might be a more frequently used technique for winning.

Nonetheless, our perception seems to be biased: we will readily believe that taller individuals have a higher status. Conventional wisdom and to some extent the scientific literature suggest that US presidential elections are won, more often than not, by the taller of the two candidates. Taller leaders are seen as stronger leaders, researchers say. Height has been particularly important in wartime, as the heights of Woodrow Wilson (5′11″) and Franklin D. Roosevelt (6′2″) suggest. In particular, during times of threat, we have a preference for having a "Big Man" as a leader. A small difference does not really matter: Barack Obama (6′1″) won the 2012 presidential election against Mitt Romney, who was an inch taller.

West Point, a leading military institution known for producing high-ranking military officers, instituted a policy in the 1950s to prevent height from influencing initial cadet ranking. This was done by assigning cadets to groups that were organized by height. This was originally done to make troops appear more uniform when marching, but it also served to prevent height discrimination. The promotions of cadets were evenly spread throughout the different groups. This constraint was later removed, as it was shown that height had only a small effect on slowing the promotions of the shortest men, giving the tallest only a slight edge in reaching the highest rank of general. While there was still some correlation, the numbers weren't in favor of keeping the height-ordered groups. To this day West Point promotions are still well spread out, with only a slight favor toward taller men. It is possible that those early height-ranked cadet groups had a lasting impact on promotions at West Point, but as time goes on the advantages of taller candidates should become more apparent if bias is really present. Height has only a slight effect in the ranking of military individuals, and the effects are small enough not to need intervention when training cadets.

Evolutionary psychology is the theoretical framework that attempts to explain how our brain evolved to provide us with a survival kit for the Stone Age and subsequently produced the culture we live in. Mark van Vagt, a Dutch evolutionary psychologist, and his colleagues have found a relationship between social rank and our perception about physical features:

- Individuals are estimated to be taller if they have higher status, obtained either via dominance or via prestige.

- Taller individuals are estimated to have higher prestige- and dominance-based statuses.
- Dominant high-status individuals are perceived as more muscular than prestigious high-status individuals.
- More muscular individuals are perceived as dominant but not necessarily as prestigious.
- As opposed to adults, primary school–aged children correlate size with dominance but not with prestige. This finding suggests that while dominance may be universally linked to a perception of increased size, the relationship between height and prestige is culturally learned.

I don't believe that the legendarily prolific Hungarian mathematician Paul Erdős was taller than 5′6″. While it is true that he was neither colonel nor president, he has still been considered the monarch of mathematics. In addition, in the country of mathematics the proclamation known from other kingdoms, "the king is dead, long live the king!" is not valid: more than 20 years after his death, Erdős is still the king of mathematics.

Social structures: hierarchical versus network organization

Hierarchies

Hierarchy is the very general organizational principle that characterizes our physical, biological, and social systems.[13] Hierarchies are structured in layers or levels. An excellent example from the field of interdisciplinary science is the evolution of complex, hierarchical human societies, which has been explored by combining the collection and analysis of traditional historical data with mathematical modeling. The hypothesis at the core of this research deals with two main governing factors: warfare and what is called *multilevel selection*, both of which have propelled human evolution for centuries.[14] According to anthropological scholars, human society evolved from small-scale, relatively egalitarian tribes to the complex social entities of advanced industrialized nations through a multilevel selection mechanism combining competition and cooperation. Historically, tribal social groupings resulted in competition for scarce resources, and tribes had incentives to act selfishly for the benefit of their group members. However, historians and anthropologists have noted that during periods of intensive competition, like wartime, tribal groups also tended toward cooperative practices. Increased cooperation among group members induces firmer social cohesion, drives technological progress (including military and organizational applications), and results in population expansion. Due to cognitive bounds that limit the number of social relationships any single person can maintain, evolutionary mechanisms have promoted demarcating social groups along cultural, linguistic, religious, and other lines, and constructing ever-larger social hierarchies that have grown to encompass societies of quite literally billions of people.

These interdisciplinary studies show that (1) both altruism (benefiting others within a community at a cost to ourselves) and hostility (toward individuals outside of our own ethnic or racial communities) are common human behaviors and (2) the intersection of the two (called *parochial altruism*)[15] led to an evolutionary mechanism that has since been well labeled with the slogan "cooperate to compete" for generating large-scale hierarchical social structures. Social scientists like to say that purely biological mechanisms cannot account alone for the formation of social hierarchies, and I am ready to accept that nowadays it is a good working hypothesis to consider biological–social coupling. As Herbert Simon suggested, problem solving was the mechanism that led to hierarchies, and the expertise of individuals and labor division was the main tool, rather than selfish gene-type competition.[16]

Some social hierarchies

Social hierarchies can be traced throughout history, and here are a few examples that range from ancient to modern.

The strict social hierarchy of the Aztecs

Aztec society was structured by social, political, and religious hierarchies. The election system ensured continuity: emperors were usually chosen from among the brothers or sons of the deceased ruler, and they were elected by a high council of four nobles who were related to the previous ruler. The nobles had many privileges, including full educations and fancier clothes. They might have held government offices, but craftsmen, or even servants, could be nobles. There was some class mobility, since servants with distinction could move up in the ranks. The commoners were farmers, artisans, merchants, and low-level priests. Slaves typically had more rights than we would often think: they had the right to form a family and even to buy their freedom. (Actually, poor and free people could sell themselves as slaves.)[17] Figure 3.1 illustrates their social structure.

Hierarchies in medieval Europe

The feudal system in medieval Europe involved a strict "pecking order": everyone from the pope to the king to the peasant knew her place in the hierarchy.[18] The king was on the top of the hierarchy and possessed maximal power in the structure, which was based on the belief that God owned the land, and the king, who ruled with God's permission, could use this land. The king granted the land to nobles in return for military services. Nobles in turn granted the land to peasants, who conducted the agricultural work and provided other services. Land and privileges were subdivided based on the hierarchy shown in Figure 3.1.

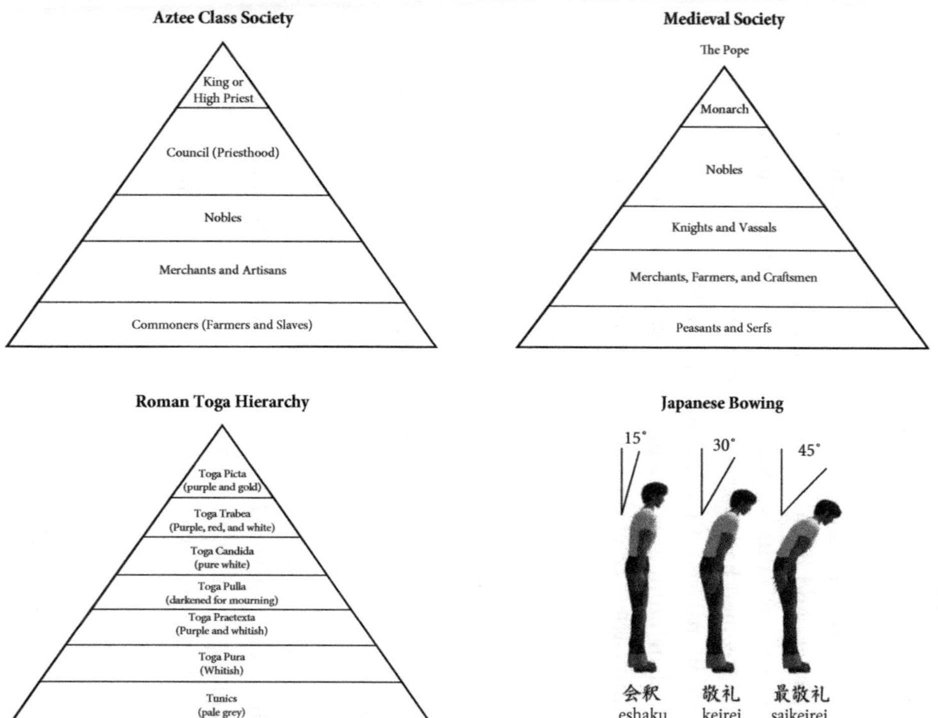

Fig. 3.1 Clockwise from top left, these figures depict the Aztec social structure, the feudal hierarchy of medieval Europe, the significance of bowing in first-century Japan (image via Wikimedia Commons), and the measure of social status indicated by toga color in ancient Rome.

Toga and rank

Roman society was strongly hierarchical. Among the *patricians* there were wealthy landowners and politicians, like consuls, senators, and judges, who could veto specific laws. Among the most famous patrician families are the families of Julia (Julius Caesar), Cornelia, Claudia, Fabia, and Valeria. The *plebeians* (the ordinary citizens who had to pay taxes) occupied common jobs like shopowners and could not participate in government. (One of Rome's most famous senators, Cicero, however, was a plebeian.) The freemen had jobs as craftsmen and traders, and while they were not enslaved, they had few rights. Slaves had no rights and worked in mining, farming, construction, and other labor-intensive industries. Over the course of many years of work slaves could save up and eventually buy their freedom. The hierarchical organization was well reflected by the types of toga worn, as shown in Figure 3.1.[19]

Symbolic actions of social status

Bowing in Japan is a clear example of how *action* can be a demonstration of relative status. Historically, bowing started during the Asuka and Nara periods (538–794), and the tradition goes back to Chinese Buddhism. In modern Japanese society, the action of bowing has been preserved, and it expresses a variety of attitudes, from thanking to apologizing to congratulating, etc. There is a hierarchy of bowing expressed by the angle of bending[20] (see Figure 3.1).

Social dominance orientation

Social dominance orientation (SDO) measures social and political attitudes. SDO is measured by responses to a series of statements using what is known as the Likert scale. A frequently used version of the Likert scale offers five possible answers (strongly disagree; disagree; neither agree nor disagree; agree; strongly agree), with numbers from one to five assigned to each of these answers. Here are some textbook examples of the statements tested in measuring SDO, against which you may check your own attitude:

- Western civilization has brought more progress than all other cultural traditions.
- Lower wages for women and ethnic minorities simply reflect lower skill and education levels.
- Patriotism is the most important qualification for a politician.
- If not executed, murderers will commit more crimes in the future.

SDO measures attitudes regarding inequality between social groups. While measurement refers to the present, it also predicts future behavior: "SDO also predicts support for group-relevant social policies that uphold the hierarchical status quo, such as support for wars of aggression, punitive criminal justice policies, the death penalty and torture, and opposition to humanitarian practices, social welfare, and affirmative action."[21]

How does our brain help us to know our social rank?

Do you remember how you spent the first days in your newest job? I would guess many of us, consciously or unconsciously, collected information about the formal and informal relationships among people. It takes longer to uncover the structure of the gossip network and the informal social hierarchies. It was easy to notice that a colleague whose office was adjacent to mine seemed to know everything about everybody on our small campus. So for me it is a viable strategy to visit him occasionally to update myself about recent (and future) local events.

Modern neuroscience has combined brain imaging devices and computational techniques to uncover some mechanisms about how our brains process information on social hierarchy.[22,23] This exciting field, known as social neuroscience, uncovers the brain regions and neural mechanisms related to reflecting ranks and dominance. Studies have shown that a brain region called the dorsolateral prefrontal cortex might play a significant role in the prevalence of employment discrimination against women or ethnic minorities, which is directly related to the conservative and hierarchy-enhancing attitudes indexed by the SDO scale.

So what? Assuming neural determinism, may an ultraconservative guy say, "It's not me, only my prefrontal cortex"? It is not easy to give a good answer. A new scientific field is emerging in the overlapping area of neuroscience and law to discuss the relationship among neural mechanisms, free will, and criminal responsibility.[24] I can't do better than leave the question open for the time being.

Network societies?

"Network" has become a buzzword. Transportation and trade networks, food webs, electric power networks, the World Wide Web and the Internet, and social networks belong now to our everyday lives. Sociologists have suggested[25,26] that society can better be seen as a complex network than a simpler, purely hierarchical structure. There were several factors that historically boosted the emergence of a network society: (1) the spirit of *open markets* promotes the absence of regulatory barriers to free economic activity; for instance, the stock markets are open because any investor can participate, the prices are the same for all players, independent of their location on the social hierarchy, and they vary based on shifts in supply and demand; (2) the spirit of the freedom-oriented political and cultural movements of the late 1960s; and (3) the revolution in the information and communication technologies. Niall

Ferguson, in his bestseller *The Square and the Tower*,[27] suggests that history may be seen as a clash between hierarchies and networks.

One example of this clash is related to the emergence of decentralized currencies, such as bitcoin. There are several pieces of green paper in my pocket, and I will trade them soon for some brownish warm liquid. Why will I be able to get coffee with these green pieces of paper? Because the US government guarantees that this paper has a nominal value. Bitcoin is a *decentralized currency*. To mine a US dollar is totally unlawful, but you can mine bitcoin. When we use a dollar we trust in the US government, whereas the value of bitcoin comes from the network of people who use it.

While in hierarchical systems it is trivial to determine who is on the top, researchers have traditionally used a variety of measures to decide who is in the center of a network community. Once a predefined centrality measure has been selected, it is possible to characterize each member of the community with a score reflecting her rank.

Figure 3.2 shows a network of 10 people.

Three possible centrality measures are shown in Table 3.1.

Degree centrality is characterized by the number of direct connections a node has. In the network shown in Figure 3.2, Sean and Bill have the greatest number of direct connections. They are "connectors" or "hubs." In contrast, Ellie has few direct connections—fewer than the average in the network. Yet, in many ways, she has one of the best locations in the network because she is between two important constituencies and plays a "broker" role in the network. Emily and Ellie have fewer connections than Jane and Lucy, yet the pattern of their direct and indirect ties allows them to access all the nodes in the network more quickly than anyone else. They enjoy many of the shortest paths to all others—they are close to everyone else.

Thus, we can say that Bill and Sean are on the top based on "degree centrality," while Sean is also closest to everybody else and therefore first in "closeness centrality." Ellie ranks first in "betweenness

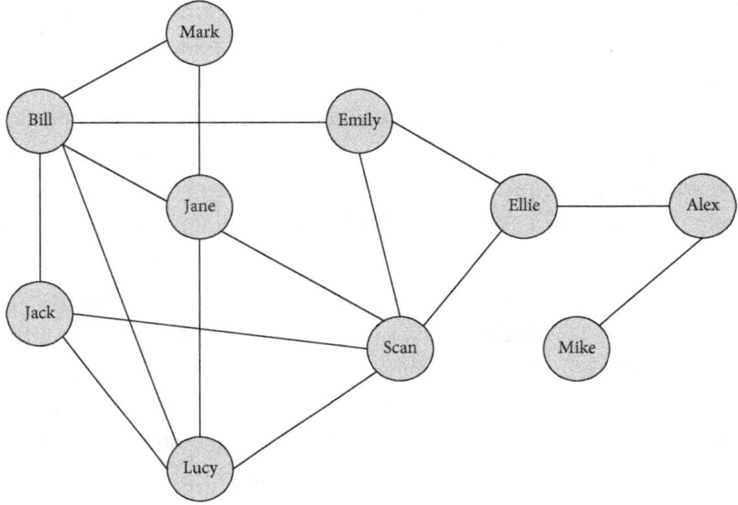

Fig. 3.2 Network of a community of 10 members. Different methods result in different rankings of the members emphasizing different features.

centrality," so she is in a position to mediate and process the most information (Table 3.2).

There are two lessons to learn. First, hierarchical organization can trivially tell us who is on the top, while network organization cannot. Second, it is possible to tell who are the "leaders" of a

Table 3.1 Ratings for the three centrality measures

Nodes	Mark	Bill	Jack	Jane	Lucy	Sean	Emily	Ellie	Alex	Mike
Degree rating	0.222	0.556	0.333	0.444	0.444	0.556	0.333	0.333	0.222	0.111
Closeness rating	0.409	0.563	0.500	0.529	0.529	0.643	0.563	0.563	0.409	0.030
Betweenness rating	0.000	0.129	0.006	0.072	0.013	0.274	0.106	0.311	0.178	0.000

Table 3.2 Rankings for the three centrality measures

Nodes	Degree Ranking	Closeness Ranking	Betweenness Ranking
Mark	4	4	8
Bill	1	2	4
Jack	3	3	5
Jane	2	3	6
Lucy	2	3	7
Sean	1	1	2
Emily	3	2	5
Ellie	3	2	1
Alex	4	4	3
Mike	5	6	8

community in a network, but different criteria give different results. A person might be in the center locally but may have less effect on distant members.

Ranking fight: democracy versus authoritarianism 2.0

According to the overused quote of Churchill, "Democracy is the worst form of government, except for all the others." The early American democracy was based on the combination of hierarchical and network structures and proved to be very efficient. While most governments operate on the *rule of law*, which prescribes that everybody—yes, everybody, "believe me," everybody, even the president—is equally subject to law, democratic societies are based on the key principle of free and fair elections.

Through the lens of the ranking game, recent concerns about the crisis of our democracy and democratic elections can be seen as a fight for dominance. Many recent books with such titles as *How Democracies Die; Against Elections: The Case for Democracy; Our Damaged Democracy: We the People Must Act; The People vs. Democracy: Why Our Freedom Is in Danger and How to Save It*[28,29,30,31] share this concern.

Democracy, more often than not, means "representative democracy" (and not "direct democracy"). A small number of self-selected people put themselves up for election. From the perspective of statistical idealism candidates should be selected randomly from the population to be really *representative*. Well, let's return to the real world!

Gerrymandering is a tool used to lawfully manipulate elections (a formal method of ranking candidates), and recently there have been partisan cases in Maryland, Wisconsin, and North Carolina[32] related to the legality and constitutionality of the strategy. Historically, the manipulation of voters has occurred through vote buying and voter intimidation.

A new chapter of manipulation started with the Cambridge Analytica scandal. My interest in ranking was initiated by a project I have been working on with a series of students and colleagues over the last 15 years to analyze patent databases, and Facebook actually has patented some of its technology. The description of US patent *US20140365577A1*, *Determining User Personality Characteristics From Social Networking System Communications And Characteristics* (owned by Facebook), states:

> A social networking system obtains linguistic data from a user's text communications on the social networking system. For example, occurrences of words in various types of communications by the user in the social networking system are determined. The

> linguistic data and non-linguistic data associated with the user are used in a trained model to predict one or more personality characteristics for the user. The inferred personality characteristics are stored in connection with the user's profile, and may be used for targeting, ranking, selecting versions of products, and various other purposes.

To build authoritarianism 2.0 the first step is to obtain (a lot of) data; the second step is the use of predictive data analysis, which creates psychometric profiles of people; and the third step is to influence people with information and disinformation. If disinformation wins, we (I mean, my grandchildren and their peers) will live in authoritarianism 2.0. I try to believe that even manipulation has its limits and will not happen.

Lessons learned: evolution and beyond

Social ranking among people certainly has biological roots. Dominance and prestige are the key mechanisms of forming social hierarchies. Hierarchical organization has proven to be efficient for communities, since it helps individuals avoid superfluous fights. Dominance is an evolutionarily older strategy and is based on coercion. Prestige is younger and is based on skills and knowledge that are generally accepted and appraised by the community. The original, strict, hierarchically organized societies adapted a good number of networks (from transportation and trade networks to electrical networks to modern communication networks). The adaptation of these networks contributed to the reinforcement of the democratic societies. But something went wrong, and during the past several years we have seen a reemergence of the ranking fight between democracy and authoritarianism.

Social ranking generally emerges as the consequence of the decisions and choices of individual people in a community. In the next chapter we will discuss the scope and limits of rational choices and also how the choices of individuals are aggregated to express the opinion of a community. We will also see that the results of the ranking process are not always unique, and we will learn how to understand the results of the ranking game.

4

Choices, games, laws, and the Web

From individual to social choices

The notion of "objective reality" refers to anything that exists independently of any conscious awareness or perceiver. By contrast, "subjective reality" is related to anything that depends on some conscious awareness or some perceiver. Objectivity is associated with concepts like reality, truth, and reliability. Objectively ranking the tallest buildings in the world is relatively easy, since it is based on verifiable facts, and we have a result that everybody will accept. (Well, we must exercise some judgment in evaluating secondary or tertiary sources of information. I have not measured the heights of the towers myself, but I am ready to accept that the information found on the Web[1] is reliable, and other webpages give the same result.) Here's the list:

1. Burj Khalifa, United Arab Emirates: 2,717 feet
2. Shanghai Tower, China: 2,073 feet
3. Makkah Royal Clock Tower, Saudi Arabia: 1,972 feet
4. Ping An International Finance Centre, China: 1,965 feet
5. Lotte World Tower, South Korea: 1,819 feet
6. One World Trade Center, United States, 1,776 feet

Ranking individuals in terms of historical influence is a more difficult task. The website Ranker gives the following ordered list of the most influential people of all time:[2] Jesus Christ, Albert Einstein, Isaac Newton, Leonardo da Vinci, Aristotle, and Muhammad.

This list is far from being objective: What about Napoleon, Hitler, Stalin, Churchill, Darwin, and countless others? The ranked list of tall buildings is objective, as it approaches what we accept as the "truth," and we feel that trying to create an objective ranking is more valuable than trying to create a subjective one. A person who argues based on a subjective viewpoint sees things only from her own perspective, which is inevitably loaded with all sorts of biases. Of course, *subjective* is not identical to *random*. When you read the list of the six most influential people, you were likely not surprised by any of the names, even if you thought the list missed somebody else. However, if you saw on this list the name Péter Érdi, you might think that the list was random, extremely subjective, or, most likely, manipulated.

Philosophers have largely given up on the attractive concept of absolute objectivity, but still mathematicians study the science of objective ranking.[3,4] However, nobody is perfect, and no individual can be fully unbiased.

How do people choose?

The myth of rationality

We like to believe that we are rational, and what is known as the *neoclassical theory* of economics is based on the assumptions that humans have fixed preferences and that these preferences are transitive. When we speak of "fixed preferences," we mean that if you prefer Key lime pie to caramel fudge cheesecake on Monday, you will do so on Tuesday too. (This book is not about healthy desserts.) Transitivity means, for example, that if, deciding on a dessert to complete your dinner at a fine-dining restaurant, you prefer Key lime pie to caramel fudge cheesecake and this cheesecake to chocolate mousse, then you will prefer the pie to the chocolate mousse as well. So, the choice of a dessert is a *rational* activity, but only in a restricted sense of the word. We would not tell somebody that they are irrational because they nonetheless choose chocolate mousse over Key lime pie.

Neoclassical economic theories are based on the concept that we are rational in the sense that during decision making, humans are concerned with maximizing their expected gain (say, pleasure or profit), which can be expressed by a *utility function*. If we want to undertake a quantitative analysis, say one that maximizes the utility function for our dessert selection, we should be able to assign numerical values to our desires to consume pie, cheesecake, or mousse. The development of *rational choice theory*[5] in social sciences made it possible to represent and solve problems of choice in a formal manner and has since served as the basis of many results in decision theory, game theory, and microeconomics.

Rational choice theory is based on absurdly simple assumptions: that more is always better, that people have full information, and that people can use this information rationally. In addition, rational choice theory assumes that people are not affected by their emotions, such as fear or envy. So, more or less, the model assumes that people are emotionless robots who don't make computational errors. In a famous publication (6,325 citations noted by Google Scholar as of June 7, 2018), the celebrated economist Milton Friedman[6] argued that it is possible to offer useful prediction tools for economists even based on these oversimplified assumptions.

So how do social decisions emerge in the society of *Homo economicus*? Before attempting to give an answer, we should clarify the differences between two types of theories: descriptive and normative. The first attempts to answer the question "How does the world work?", while the second is interested in the problem of "How *should* the world work?" Responses to the second question have been given in both mathematics and moral philosophy, and I would like to believe that the best responses come from some combination of the two disciplines. In an ideal society of rational people, what is known as the Pareto optimality state corresponds to a state of resource allocation in which it is impossible to improve any single individual's utility without making at least one other individual worse off. Welfare economics, a subdiscipline of economics more generally, aims to determine the state that creates the highest overall level of social satisfaction among its members. Technically, the task is to give a complete and transitive ranking of all social alternatives.

As always in the case of ranking and rating problems, there are a number of possibilities. First, what economists call an *ordinal utility function* allows an individual to rank all possible "states" in an ordered list. The output of the ordinal utility function indicates that an individual prefers possible state *X* to possible state *Y*, but such a function does not account for the *strength* of preferences. Second, a *cardinal utility function* assigns a number to characterize the attractiveness of any state, so it is possible to express the magnitude of how much an individual prefers *X* to *Y*. Here is a recurring problem: How do we assign numerical quantities to qualities? In a very, very reduced sense, the world is nothing but a set of products to buy, so one way to assign a numerical quality to the utility of a good is to ask what price people are willing to pay for that good. If you are ready to pay $27,000 for a Toyota Camry Hybrid, you can say that this product has 27,000 utils (the abstract unit of utility). But this is a very naïve picture.

The next problem we face concerns the aggregation of individual preferences: How do we construct a social welfare function (SWF) from individual utility functions? One way might be to define a SWF as the sum of each individual utility. In this case, the goal of maximizing the SWF means maximizing the individual incomes, but this algorithm totally neglects any concern for income *distribution*. A very skewed distribution, where a small portion of people (say, the top 1 percent) owns the majority of wealth, may maximize the SWF when defined this way. (Another option is to define SWF based on the *average* of individual utilities.)

John Rawls (1921–2002), a leading political and moral philosopher of the last century, identified classical and average utilitarianism as "the principal opponent" of justice and suggested defining a SWF based on the utility of the *worst-off* individual. So, maximizing SWF would mean maximizing the income of the poorest person in society without taking into account the income of other individuals.[7]

Amartya Sen, the legendary Indian economist, proposed an SWF that penalizes economic inequality.[8] Sen used both income measures and the Gini coefficient (G), a quantitative measure of economic inequality, in formulating his SWF. G is zero in the case of perfect equality (i.e., everyone has exactly the same level of income), and G is one in the case of extreme inequality (i.e., one person receives all the income and others receive none). Sen's SWF is defined as the average per capita income of a country, multiplied by the number $1 - G$. A good normative theory can be applied to the real world, and we will return to Sen's SWF as we discuss ranking countries in Chapter 6.

Against the myth I: bounded rationality

While probably not even the most dogmatic economist truly believes that the "hyper-rational," utility-maximizing agent is a plausible model for describing human behavior, and a variety of criticisms have been leveled against rational choice theory, a more realistic paradigm is gradually emerging. Herbert Simon (1916–2001), who worked far from the mainstream, somewhat unexpectedly received the Nobel Prize in economics in 1978 for introducing and propagating the concept of *bounded rationality*. Bounded rationality rejects the need for a perfect solution and instead acknowledges that a satisfactory, even suboptimal, solution is sometimes good enough. He coined the term *satisficing*, combining the words *satisfy* and *suffice*, to describe the process of making a "good enough" decision. We have to accept that our ability to make decisions is limited by a number of constraints, such as the complexity of a problem, the available resources (most importantly time and money), available information, our cognitive skills, our values, the influence of our feelings, and countless other factors. As I learned on the soccer field in my childhood, "there is no better position than a good one."

There is a nice mathematical problem, officially known as the "optimal stopping problem," that helps explain when to stop dating and choose a long-term mate. It serves as a good example of satisficing. The problem is analogous to the problem of an employer trying to find a suitable new office manager from a range of applicants, so it is also known as "the secretary problem." Let's assume that you have a number of possible mates. I don't want to specify a number, but if your name is not Don Giovanni, the number is much less than 1,003 ("Ma in Spagna son gia mille e tre"). As opposed to Don Giovanni, suppose you have one relationship at any given time. You should decide whether or not he or she is "the one." More often than not, you can't go back to one of the people you have rejected earlier. You might make two types of mistakes: (1) you may decide to settle down early and possibly wonder in the future whether you missed the chance to meet the real king or queen of your life or (2) you may wait too long to commit, and all the good ones might be gone. So, here is the big question: When should you stop? Math gives an answer for the magic number: 37 percent.[9] You have the highest chance of finding Ms. or Mr. Right if you date and reject the first 37 percent of your potential mates. The rule has a second part: pick the next person who is better than anyone you have ever dated earlier. (Yes, the algorithm does not guarantee that you will not reject a wonderful option, so you must balance the risk of stopping too soon against the risk of stopping too late.)

While this gives a somewhat trivial example of how bounded rationality impacts our romantic relationships, Bryan Jones, who has been serving as the J. J. "Jake" Pickle Regent's Chair of Congressional

Studies in the Department of Government at the University of Texas at Austin, has provided insight into how bounded rationality impacts political decision making:

> As Herbert Simon . . . notes, *Homo politicus* is not irrational. He seems to behave purposefully, adopting strategies that are relevant to general goals, given the limits of cognitive capacity and the complexity of the political world. But these facets make it impossible to maximize and often inappropriate to maximize. *Homo politicus* seems to Simon to operate according to the model of bounded rationality, that is, adopting means that are relevant to goals within environmental and cognitive processing limits.

Against the myth II: from rational choice to behavioral economics

Behavioral economics has evolved rapidly in the last 20 years and has risen to challenge the rational choice theory underpinning a substantial portion of economic theory. The driving force of this evolution came from the recognition that assumptions with greater psychological plausibility generate theories with greater explanatory power. Empirical observations combined with experimental studies helped researchers Daniel Kahneman and Amos Tversky (1937–1996) discover the phenomenon of *cognitive bias* that systematically characterizes our thoughts. Cognitive bias makes us "predictably irrational," to use a fashionable expression. It frequently blocks us from making rational decisions, even when we try our best. But what is rational behavior? Is it behaving in favor of our narrow economic interests? A counterexample to this definition of "rational" is the Ultimatum Game.

In the Ultimatum Game, there are two players—the proposer and the responder—who have to agree on how to split a certain amount of money. The proposer makes an offer. The responder has two possibilities: accept or reject. If she accepts, the deal is done. If she rejects, neither player gets a single penny. Rationality would require the responder to accept any positive offer, even the smallest one. In most cases, the proposer would obtain the overwhelming majority of the entire sum. However, studies across various cultures have shown that responders tend to reject offers below 30 percent of the sum in question. Still, we may say the expected utility concept works if we should take into account the psychological benefit of being able to reject an offer just to penalize the miserly proposer.[10,11,12]

Sources of cognitive bias

We can give a list of factors contributing to the deviation from the assumed behavior of *Homo economicus* based on observations and experiments:

The phenomenon called *availability bias* can be identified when we overestimate the probability of an event occurring because a similar event either has happened recently or had a significant emotional impact in the past. (I remember a family vacation in Dubrovnik, when my then-14-year-old son, an excellent swimmer, refused to bathe in the Adriatic Sea, overestimating the likelihood of a shark attack, since a few days earlier he had watched a shark movie. I cannot remember what movie it was, and when I tried to find it on the Web, this listicle came forward: "13 shark movies that will make you avoid the water forever."[13] If you plan a family vacation at the seaside, you may not wish to open the website given in the notes.

Hindsight bias is a mental error that occurs when we falsely believe we have predicted an outcome. The expression "hindsight is 20/20" refers to optometric studies that measure and rate visual acuity. A measurement of 20/20 is considered perfect vision, meaning you can read stock quotes in the newspaper or numbers in the telephone book, for example. (When was the last time you used a phone book?) We are often ready to claim "I knew it!" The fall of the Berlin Wall on November 9, 1989, and the peaceful unification of Germany were not predictable. It was not unimaginable that a Tiananmen-style event might have occurred, as this tragedy had happened just a few months earlier in Beijing. I like Henri Bergson's expression "the illusions of retrospective determinism," or the similar phrase *a posteriori wisdom*. Despite the unpredictability, many people claimed they had foreseen the fall of the wall. Another example of a posteriori wisdom is our relationship to Brexit. No one knew the outcome of the referendum in 2016 before it happened, not even David Cameron, who initiated it. The day before people went to polls was the busiest political betting day in history, and bookmakers largely bet on Britain staying in the European Union.[14] The data used by bookmakers, however, were controversial: 69 of all the *money* they took was for "remain," but 69 percent of all the *individual* bets were for "leave."[15] While it seems to be true that the majority of voters have chosen to increase uncertainty, I am sure they were not thinking of the long-term implications of such a decision. Here are some news headlines from almost two years after the referendum (today, June 14, 2018): "Foie gras imports may be banned after Brexit, UK minister suggests"; "Security row over EU Galileo satellite project as Britain is shut out"; "Bankers to ask May why they should stay in London after Brexit." While Brexit was not at all inevitable, it happened. Hindsight bias helps people accept their decisions, even when such decisions are made against their best financial interests.

Once we learn the numerical values of certain things, we feel that they are "just right." This is called the *anchoring effect*. In a centrally planned economy, prices are determined by the state. The price of a kilogram of bread was 3.60 Hungarian forint for decades, since it was a political decision to keep the price of the bread constant. Everybody from my age group remembers that this is the correct price of bread. I also learned in my childhood that Jesse Owens's world record in the long jump was 813 cm, and it stood for 25 years. It was slowly overshadowed in my mind by another mythical world record, 890 cm, which Bob Beamon achieved in the long jump in the Mexico City Olympics in 1968.

Confirmation bias occurs because our brain is wired to prefer things that conform with our preconceived notions. Over the course of many years, we acquire a slowly evolving system of beliefs. By using a somewhat more technical expression, we have an internal mental model about the external world. Our mind tends to incorporate new incoming information in our mental model in a way that maintains overall coherence. Francis Bacon described this centuries ago:

> The human understanding when it has once adopted an opinion draws all things else to support and agree with it. And though there be a greater number and weight of instances to be found on the other side, yet these it either neglects and despises, or else by some distinction sets aside and rejects; in order that by this great and pernicious predetermination the authority of its former conclusions may remain inviolate.[16]

The Central European intelligentsia like to cite (with some irony) the German idealist philosopher Johann Gottlieb Fichte (1762–1814), who said (or might have said): "If theory conflicts with the facts, so much the worse for the facts." The expression "fake news" became very popular in 2017. It is unfortunate, but better to know than to be ignorant of the fact, that "we" (and not only "they") can be victims of fake news.

I turn again to some results coming from the fantastic new field of social neuroscience, which has started to uncover the specific neural mechanisms behind confirmation bias.[17] If we encounter a statement that does not conform to our existing views, an obvious conflict is generated. Political beliefs are an important constituent of our social identity, and behavioral data show that we are more flexible in accepting and modifying our beliefs in response to conflicting nonpolitical statements than conflicting political ones.

Brain imaging studies have shown that those people who showed the largest resistance to changes in their political beliefs in response to counterevidence demonstrated an increased activity in the amygdala and the insular cortex, the brain regions involved in fear and emotional response. Positions about gun control and abortion, which are especially central to American political debates, seem to be particularly stable. If we feel ourselves threatened, anxious, or otherwise generally emotionally attacked, we don't really like to change our mind. It is very important to accept the interaction between our rational, cognitive system and our emotional system. We do not accept new things easily. I find a silver lining in these studies: maybe we cannot be manipulated too easily.

I teach a class called "Introduction to Complex Systems" each winter, and a component of the class is a group project where the students make computer simulations of some biological or social problem. Generally there are four students in the group, and they work together for about seven weeks. At the end of the term, sometimes I ask them to anonymously report their contribution to the group's work in terms of a percentage. You will not be surprised that the sum of the percentages falls between 130 and 170 percent, which is significantly higher than 100 percent. Please note, they don't get any credit for the number they report. It is simply the case that people overestimate their own contribution to group endeavors. (It brings to mind Garrison Keillor's closing on each episode of the long-broadcast radio program, *A Prairie Home Companion*: "Well, that's the news from Lake Wobegon, where all the women are strong, all the men are good looking, and all the children are above average.") I don't really blame the wishful thinking related to *egocentric bias*, since it is a self-defense mechanism. Even though the divorce rate is now close to 50 percent, people don't think that the likelihood of their marriage ending in divorce is 50 percent. We apply wishful thinking to events we cannot fully control. It is not necessarily very bad to believe we are less likely to be at risk for negative events, such as developing cancer, getting divorced, or having an automobile accident.

Loss aversion bias is the belief that losses are bigger than similar-sized gains, as the now-celebrated "prospect theory," developed by Daniel Kahneman and Amos Tversky, states: people tend to fear a loss twice as much as they are likely to welcome an equivalent gain. When you read "twice," it is just an approximate number: the sadness of the loss of $50 is approximately equivalent to the happiness of gaining $100. You may ask how it is that we can measure sadness. While we are wired with loss aversion, we should know that it influences our decision. Here is a famous example:[18]

> Imagine a cancer patient with six months to live. Their doctor comes to her and says, "There's a new treatment! We'd need to do this right away, and if it's successful you'll be cured, but there's a 10 percent chance that you'll die during the treatment."
>
> Then, on the other side of the country, another cancer patient with the same time to live has their doctor visit them: "There's a new treatment! We'd need to do this right away. If it's successful you'll be cured, and there's a 90 percent chance that you will survive the treatment!"
>
> The second cancer patient is much, much more likely to take up the treatment than the first. Yet, the two statements are identical. The chances of dying remain at 10 percent and the chances of living remain at 90 percent, but one statement invokes our fear of loss while the other does not, and the stakes don't come any higher than our lives, do they?
>
> This is very important because it means that we need to think about the way we communicate with other people; when we seek action, are we focusing too much on what could be lost rather than what could be gained?

This is an example of what is called the *framing effect*. People tend to accept an option when it is framed positively, and reject it when framed negatively.

Choice as a source of happiness and misery

Barry Schwartz's influential 2004 book *The Paradox of Choice: Why More Is Less* was motivated by Herbert Simon's concept of "bounded rationality" and describes the conflict between the "maximizers" (those who always search for the best possible choice) and the "satisficers" (those who feel that "good enough" really is good). While logic might suggest that having more options makes us happier, it is not necessarily true. How many options of toothpastes, insurance policies, colleges, long-term partners, cereals, retirement plans, cellphones, vacation plans, or TV channels do we need? There are cognitive limits to our comparative evaluations of too many things, events, or other objects. Maximizers might, as a result, have the feeling that they chose a suboptimal option. They might blame themselves for making insufficiently good decisions, and the feeling may make them unhappy or even depressed. Social media makes inescapable the overabundance of everything, and, as a result, it has dramatically amplified an ever-present feeling that is called the "fear of missing out," or, as we all have come to know it, FOMO. Recent social psychological studies have provided data mostly related to adolescents and college students, but even the title of a single paper is illuminating: "'I don't want to miss a thing': adolescents' fear of missing out and its relationship to adolescents' social needs, Facebook use, and Facebook-related stress."[19,20] It remains to be seen whether or not we will be able to educate the next generation to increase their degree of internal autonomy and decide among the seemingly infinite options before them.

However, there are some recipes for avoiding the overwhelming onslaught of options:[21]

- Consciously restrict your options. It might be enough to visit two stores in a mall when shopping for clothing.

- Learn to stop when you meet "good enough."
- Don't worry about what you're missing.
- Don't expect too much, and you won't be disappointed.

Marry Mr. Goodenough!

We need more data to justify the hypothesis (and we can't do anything better in the age of data deluge than to believe in the power of collecting and processing data, but I can already hear the critical voices protesting) that people living in long-term relationships are happier than the singletons.

In any case, in her provocative bestseller *Mr. Good Enough: The Case for Choosing a Real Man Over Holding Out for Mr. Perfect*, Lori Gottlieb argues that marrying a guy who satisfices is better than to waiting forever for Mr. Right. She believes that it is not a good idea to have unreasonably high expectations about the features of one's dream guy. It is not difficult to prepare a fixed list of several dozen characteristics you may be seeking, from hobbies to eye color. To make things more difficult, even when we have a list, the elements are not equally important. What has more weight for you, a sense of humor or financial stability? (My choice is the first, but this is for a different story.)

Maximizers have a fixed list, and they are probably able to assign specific weights to the individual features of their dream guy. They are also able to rate the real-world candidates. If the features of two objects (or subjects) are compared, the question is whether or not they are "sufficiently close" to each other. By adopting a somewhat more technical terminology, the question is whether or not the deviation is smaller or larger than a predefined threshold. If it is smaller, the real-world candidate is "good enough." The advice that Gottlieb gives is that at a certain age, it is worthwhile to increase the threshold, so that you may let pass and eventually marry

Mr. Goodenough. I will return to Mr. Goodenough when I discuss dating algorithms.

From human fallibility to being nudged

Amos Tversky, Daniel Kahneman, and Richard Thaler not only revolutionized behavioral economics, but they all also wrote bestselling books. One major takeaway of their writing is that we humans are evolutionarily wired to make errors in judgment (including ranking). We need a nudge to make decisions, even when they serve our own best interests (say, the decision to choose more healthy foods and, in my case, less Hungarian and Spanish sausage). The "nudge"[22] is a psychological mechanism aimed at influencing choices in a positive direction. Nudge generally helps to focus our attention toward specific aspects of a problem. I belong to that camp, however, that believes that nudging is value neutral and may, in fact, be used to manipulate people to act in service of negative goals.[23]

Should we accept the fact that politicians are now using the nudging technique? A good number of governments now have a team of behavioral scientists with the intention of improving the efficiency of policymaking by "nudging" their citizens. More precisely, they adopt indirect mechanisms of modifying behavioral choice as opposed to implementing direct regulations and laws. Some examples from the last several years have included a push to increase the numbers of organ donors in France and in the United Kingdom, to prevent expensive missed doctor appointments in the United Kingdom, and to boost voter turnout in elections by the Obama administration.

The behavioral economists' approach thus enhances the rational choice model, and understanding our own fallibility can help us make better choices.

Social choice

How do we aggregate individual opinions, preferences, or votes to form a collective decision in the real world? It is well known that in ancient Greece male citizens voted for their leaders, and Athens is often hailed as the earliest democracy. Voting itself was conducted by a show of hands, and the winner was declared by officials based on a visual estimate of which candidate received a majority of hands. While most European medieval political systems contained some electoral element, it wasn't until the Enlightenment and the burgeoning age of rationalism that democracy and various means of social choice became the subject of inquiry.

Nicolas de Caritat (1743–1794), often known as the Marquis de Condorcet, pioneered a particular voting system, called *pairwise majority voting*, that has remained influential even in contemporary voting studies and systems. Condorcet analyzed the behavior of juries and developed his celebrated *jury theorem* from these studies. As always, when mathematical models are used for social phenomena, we should carefully discuss the assumptions underlying these models. In this case, assuming that each member of a jury has an equal and independent chance (which is better than random [i.e., greater than 50 percent] but worse than perfect [less than 100 percent]) of making the correct conclusion, the jury theorem holds that increasing the number of members of the jury increases the probability of the group as a whole making the correct decision. Importantly, the relevance of the jury theorem is restricted to situations in which there really is a *correct* decision. It works, for example, when the members of a jury should decide whether or not a defendant is guilty. Consequently, under certain conditions, majority rules is appropriate at "tracking the truth." Of course, in real life the opinions of the voters are not independent of one another. In addition, the theorem cannot be applied to situations in which there is no "objective truth," but only individual preferences. This is the situation we encounter when we must choose among political candidates.

Condorcet's paradox is the term used for his second insight. Condorcet realized that even when individual preferences are "rational" (i.e., transitive), the resulting collective decision might be "irrational" (i.e., intransitive). For illustration, let's assume we have three voters (I, II, and III) and three candidates (A, B, and C). The voters' individual preferences are:

Voter I: A > B > C
Voter II: B > C > A
Voter III: C > A > B

When these preferences are broken down into pairwise comparisons, we obtain the following:

A vs. B: 2–1
B vs. C: 2–1
C vs. A: 2–1

So, the preference ranking for majority gives A > B > C > A > B > C > A > . . ., which is called the *Condorcet cycle*. The Condorcet paradox has been studied by a large number of people, from the perspective of both its practical role in voting and its theoretical role in mathematics.[24]

Where the practical element is concerned, electoral systems are critical means of collective decision making, and their essence boils down to ranking political candidates. Sometimes only the winner matters (as in elections for president or prime minister), but in other cases, everybody who appears on a ranked list higher than a threshold is considered a "winner" (as in some elections for members of a parliament or a board). No one has yet been able to determine the single best electoral system, and the legendary economist Kenneth Arrow (1921–2017) published his famous *impossibility theorem* in 1950 (for which he received a Nobel Prize in 1972), which showed that when voters rank candidates, some failures may occur. Arrow's studies and the subsequent work of scores of economists and mathematicians have generated debates and comparative mathematical analyses about voting systems.

Voting appears to be a relatively simple endeavor—we go to the polls, we select our preferred candidate, and the individual with the most votes wins. This is known as a first-past-the-post, or plurality, voting system, but it is just one of many possible voting systems. Our choice of a voting system, as shown by Arrow's impossibility theorem, carries vast consequences for the results of an election. Arrow's impossibility theorem requires that when we aggregate the individual preferences of voters, we meet some standards of fairness that, the theorem shows, are ultimately impossible to meet in every situation. First, there should be no individual who acts as a dictator and always has the decisive vote concerning the results of an election. Second, if all individuals prefer a particular alternative, the final outcome of the vote should reflect that preference, meaning that if all voters individually prefer candidate A to candidate B, the aggregate result should demonstrate preference for candidate A over candidate B. Third, the results should return a single ranking—there should be no ties. And finally, voters, in choosing among candidates, should consider only pairs of candidates, ignoring independent alternatives to the two candidates under consideration in each pairwise situation.[25] Condorcet's paradox is one such electoral result that might violate the criteria of fairness listed by Arrow's impossibility theorem, specifically the criterion of universality, which requires that an election produce a definitive ranking of candidates. Arrow's impossibility theorem does not suggest that every result of a system of voting is always in violation of the fairness criteria, but it does suggest that in every conceivable voting system, it is possible that votes will be cast in such a way as to violate at least one of the fairness criteria. The significance of this theorem depends on its application—as Marianne Freiberger of *+Plus Magazine* points out, whether or not the result is significant depends on how likely it is that a particular fairness criterion will be violated, and some criteria, like non-dictatorship, might be more important than others.[26] Nonetheless, Arrow's impossibility theorem shows that it is impossible to construct a perfect voting system.

Voting systems are also subject to particular kinds of manipulation depending on the method used to aggregate individual choices. If, for example, we take seriously the criterion of the irrelevance of independent alternatives required by Arrow's impossibility theorem, and we structure a voting system as a series of pairwise comparisons between candidates, the initial round of pairings can have a decisive impact on the results of the election. According to John Barrow, for instance, if we want to rig an election, using a series of pairwise comparisons, we can pit stronger candidates against one another in the earlier rounds, introducing our favored candidate only at the last minute so that she can win.[27]

Thus, we see that the structure we choose for the process of aggregating individual votes is decisive for the fairness and outcome of a particular election.

Despite what Arrow's impossibility theorem would suggest about the viability of ranked-choice voting methods, alternatives to the standard first-past-the-post voting system have been implemented in some jurisdictions across the United States. Most recently, Maine voted to continue implementing ranked-choice voting in the 2018 midterm elections, despite the state legislature's request to delay implementation of the method. Maine's method asks voters to submit a ballot that ranks candidates in order of preference, and if no candidate secures a majority of first-place votes in the first round of voting, it automatically triggers a runoff election in which one candidate is eliminated in each round and votes are redistributed according to voters' rankings until a winner is determined.[28,29]

There is an obvious disconnect between Arrow's impossibility theorem and real-world voting systems: although the theorem tells us that it is impossible to construct a fair system of social decision making, we continue to implement voting systems across the world. Because of this tension between theory and practice, challenges have been mounted to the theorem. Most notably, Michel Balinski and Rida Laraki's 2010 book[30] proposes an electoral system that addresses and overcomes some of the limitations of Arrow's impossibility theorem. They point out that Arrow's work made several key assumptions and omissions: it assumed that voters would create an ordered list of preferences, it ignored the strategic aspects of voting, and it refused to consider how one's voting preferences are informed by those of one's peers. These assumptions and omissions, they argue, have created a paradigm in social choice theory that "hypothesizes a faulty model of reality to produce an inconsistent theory."[31] Drawing on accepted measures of judgment for things like wine classification, diving and figure skating competitions, and assigning grades in a classroom, Balinski and Laraki have proposed a method of aggregating individual preferences that they call *majority judgment*. Majority judgment relies on measurement to determine rankings, as opposed to creating rankings by themselves. They propose a common language for structuring input preferences and utilities that generates electoral decisions by taking into account the *median* grade assigned to each candidate and using this median as a measure of comparison. Majority judgment works like this. Suppose that candidates are to be rated on a scale from 1 to 10, with 10 being the best possible score and 1 being the worst; if there are five voters in our scenario, two candidates may end up with score sets that look like [3, 6, 7, 7, 9] and [4, 5, 6, 7, 10]. The majority grade of each candidate is the grade that lies directly in the middle—in this case, our first candidate has a majority grade of 7, and our second candidate has a majority grade of 6. Repeated for each candidate, these majority grades can then determine an ordering of the candidates in comparison to one another, and the candidate with the highest majority grade wins. Balinski and Laraki note that this system of aggregating preferences has a key advantage: namely, a candidate's majority grade (call it α) is the highest grade approved by a majority of the voters—at least 50 percent of the voters give that candidate a score of α or higher. From here, Balinski and Laraki extrapolate their method and propose solutions for various tie-breaking situations and methods for ensuring that the system is gamed in a minimal way. But they are clear that this system overcomes the limitations of Arrow's impossibility theorem and produces a realistic method for creating social decisions.

The only thing we know for certain comes from Churchill (and he also inherited this knowledge from the past): "Many forms of Government have been tried, and will be tried in this world of sin and woe. No one pretends that democracy is perfect or all-wise. Indeed it has been said that democracy is the worst form of Government except for all those other forms that have been tried from time to time."

Rock, paper, scissors: games and laws

The hand (and mental) game called rock, paper, scissors (RPS) illustrates the difficulties of ranking even three objects in terms of their relative strength. Since rock crushes scissors, paper covers rock, and scissors cuts paper, the ordering is not transitive but *circular*. The game is played between two players, and nonrandom behaviors in opponents may be exploited to allow players to win more frequently than chance would otherwise suggest they should.

Different versions of the game date back to the Han dynasty of ancient China, and similar finger positions were also used in Japan (to encode a tiger, the village chief, and the village chief's mother: the tiger is beaten by the village chief; the village chief's mother beats the chief but is beaten by the tiger) and in Indonesia (to encode earwig, man, and elephant: the earwig drives the elephant insane; the human crushes the earwig; the elephant crushes the human).

Technically RPS is a zero–sum game (which means that one person's loss is equal to another person's gain). The natural question arises whether there is any winning strategy, but the answer cannot be expressed by a single word. If you play against a truly random algorithm, there is no way to generate any advantage. Human players, however, don't select strategies randomly. There is some variation regarding how people with different levels of experience play the game, and experienced players might try to identify patterns in their opponent's choices and exploit these observations in order to develop a winning strategy. There are some rules of thumb that can be observed, such as "winners tend to stick with the same action" or "losers change their strategy and move to the next action," often moving clockwise (R to P to S) in the sequence, since rock is the most aggressive mode and scissors is more aggressive than paper. There are other psychological observations (e.g., people use the same move twice in a row, but rarely three times) and many similar patterns can be exploited.

Why is this game interesting? The single reason is that a mathematical feature called *transitivity* is violated. We introduced the concept of transitivity earlier in the chapter, but here we shall delve a little further. A simple example of transitivity is given by the understanding that A > B and B > C implies A > C (here the symbol > means "is greater than"). The violation of transitivity leads to a cycle, in which we are not in a position to generate a ranked list.

One popular five-weapon extension of RPS is "rock, paper, scissors, Spock, lizard," since "scissors cuts paper, paper covers rock, rock crushes lizard, lizard poisons Spock, Spock smashes scissors, scissors decapitates lizard, lizard eats paper, paper disproves Spock, Spock vaporizes rock, and as it always has, rock crushes scissors," which was introduced in the American TV series *The Big Bang Theory*.

Cycling in the legal system: from the Talmud to modern times

Mechanisms leading to cyclic dominance (i.e., situations when a ranking cannot be generated) might have beneficial features. They play a crucial role in the maintenance of biodiversity in ecological systems, as in the parasite–grass–forb systems, among others. But let's make the short jump from parasites to politicians!

It is well known that the founding fathers of the United States created a system of checks and balances so that no one branch would be more powerful than another. The whole governmental system reminds me somewhat of the RPS game, but of course it is more complicated, since the pairwise comparison among the executive, legislative, and judicial branches might lead to two different results (Figure 4.1). The message is that the US governmental system was intentionally constructed to violate transitivity, since the goal was to avoid any ordered ranking among the three branches.

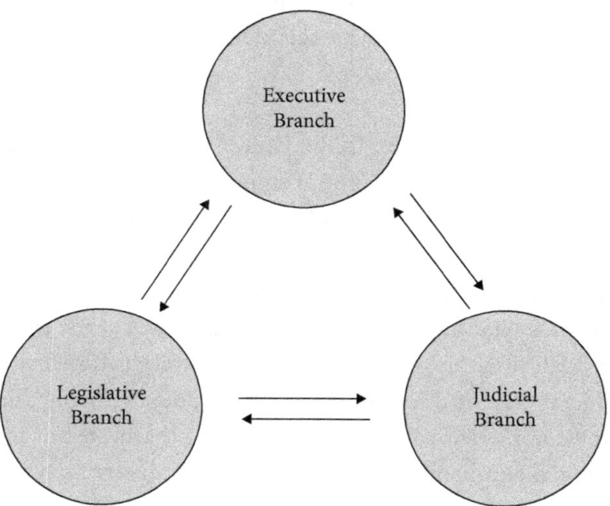

Fig. 4.1 The executive branch enforces the law. The legislative branch makes the law. The judicial branch evaluates the law.

I asked Bryan Jones, whom I mentioned earlier in the chapter in our discussion of *Homo politicus*, whether or not the US governmental system was intentionally constructed in this manner, and I feel it useful to quote his answer:

> The cycling issue for American Government is a good description on the surface, but it has problems if you get too deep. For example, there is nothing in the Constitution concerning judicial review of statutes passed by Congress and signed by the President. It was actually asserted by the court in the case of *Marbury v. Madison* (1803), and was not used again for more than 50 years. The judiciary was designed to be a weaker branch, and its major constitutional function was in the federal system––to enforce the supremacy clause that federal law was supreme if it conflicted with state law. It is true, however, that the interpretation of statues AND presidential decrees was probably implied.
>
> In any case, you might state that the US system has evolved toward an intransitive structure. But that intransitivity is limited. On the other hand, it could be a good instructional device for the general reader.

While I did not grow up in the United States, I learned that the case of *Marbury v. Madison* (1803) enhanced the power of Supreme Court, and overall, it was a good thing. (We cannot look at everything in the light of actual political events, but as I am working on this paragraph, it was recently announced that Associate Supreme Court Justice Anthony Kennedy would step down at the end of July [so President Trump gets a second opportunity to fundamentally alter the nation's top court for decades]. As I am finalizing the manuscript, he got it. We all know the details of the Brett Kavanaugh's tragicomical confirmation process.) While transitivity is a fundamental requirement for consistency, legal systems, especially when composed of various agencies, may encounter nontransitive cycles.

While the judiciary was designed to be a weaker branch, there is a much older example where even rules regarding ranking were themselves ranked. Rabbinic literature discusses inconsistent laws, like those regarding the precedence of private offerings in the Jerusalem temple. In a paper about transitivity in the Talmud,[32] scholars Shlomo Naeh and Uzi Segal discuss the relationship among rules:

- Rule 1 ranks according to type of animals, "cattle precede birds," and another rule gives rankings based on ritual functions.
- Rule 2 uses ritual function as the basis of ranking. Within each type, offerings are ranked; for example, "sin-offerings" precede "burnt-offerings."
- Another rabbinical text prescribes Rule 3: "Be it a bird sin-offering with a bird burnt-offering, be it a bird sin-offering with a cattle burnt-offering or a cattle sin-offering with a cattle burnt-offering——sin-offerings always precede burnt-offerings that are brought with them."

The third rule does not take into account the types of animal being offered, only the function of the offering. Different interpretations exist regarding how to solve the possible conflicts generated by these rules, and I leave it to the reader to study the details in the paper titled "Ranking ranking rules."[33]

The more general message is that nontransitive cycles may emerge in moral, religious, and legal systems, and it is necessary to develop pragmatic solutions to these dilemmas. Thus, the ranking of ranking rules has proven to be such a solution from the Talmud to the US governmental system.

While not every ranking problem can be solved by finding good algorithms, we now have a celebrated success story about the ranking of the World Wide Web.

The fortune-making ranking algorithm

Ranking the Web

Google is Google is Google is Google. The company's social success is reflected in its usage as a verb: "to google" or search for information on the World Wide Web using the search engine Google. The World Wide Web is a collection of webpages connected by links, while the Internet is a system of interconnected computers. The amount of information on the Web started to increase dramatically around 1993, so we users needed help navigating quickly and efficiently on the Web. So, you give a query to Google: for example, you Google "ranking the web," and Google returns a ranked list of more than 500,000 items. The first page of the results can be seen in Figure 4.2.

The list of search results was created by an algorithm. The reader will recall that an algorithm is nothing but a recipe for preparing meals—that is, a finite list of instructions. Google cofounders Sergey Brin and Larry Page made their fortunate by creating an algorithm, called PageRank, to rank websites based on their relevance. (I am not sure why it is not called the Brin–Page algorithm.) Some variation of this algorithm produced the result in Figure 4.2. There were some search engines prior to Google, but Google was much better because the new algorithm was able to answer *two* questions about each page in response to a search query: (1) How relevant is the page to a specific query? and (2) How important is a relevant page compared to other relevant pages? (Not all citations have the same weight: a link from an important site counts more.) The book *Google's PageRank and Beyond: The Science of Search Engine Rankings* by Amy N. Langville and Carl D. Meyer studies the mathematics of the operation of search engines.

About 507,000 results (0.39 seconds)

[PDF] Ranking the Web Frontier - Kevin McCurley
www.mccurley.org/papers/1p309.pdf ▾
by N Eiron - Cited by 288 - Related articles
Ranking the Web Frontier. Nadav Eiron, Kevin S. McCurley, John A. Tomlin. IBM Almaden Research Center. ABSTRACT. The celebrated PageRank algorithm ...

Ranking the web frontier - Digital Object Identifier (DOI)
doi.org/10.1145/988672.988714 ▾
Ranking the web frontier. Published by ACM 2004 Article. Bibliometrics Data Bibliometrics. · Citation Count: 65 · Downloads (cumulative): 1,571 · Downloads (12 ...

[PDF] Ranking the Web Frontier - (CUI) - UNIGE
cui.unige.ch/tcs/cours/algoweb/2005/articles/1p309.pdf ▾
by N Eiron - Cited by 288 - Related articles
Ranking the Web Frontier. Nadav Eiron, Kevin S. McCurley, John A. Tomlin. IBM Almaden Research Center. ABSTRACT. The celebrated PageRank algorithm ...

[PPT] Ranking the Web - Dipartimento di Informatica
www.di.unipi.it/~gulli/Presentazioni/Ranking%20the%20Web.ppt ▾
Ranking the Web. Gianna M. Del Corso Antonio Gullí. Dipartimento Informatica, Pisa. IIT-CNR, Pisa. Overview. Web Statistics; Some Web Ranking Algorithms ...

A survey of approaches for ranking on the web of data | SpringerLink
https://link.springer.com/article/10.1007/s10791-014-9240-0
by AJ Roa-Valverde - 2014 - Cited by 13 - Related articles
Abstract. Ranking information resources is a task that usually happens within more complex workflows and that typically occurs in any form of information ...

Ranking the Web With Radical Transparency | Linux.com | The source ...
https://www.linux.com/news/ranking-web-radical-transparency ▾
Oct 20, 2016 - Ranking every URL on the web in a transparent and reproducible way is a core concept of the Common Search project, says Sylvain Zimmer, ...

Ranking the Web with Spark - SlideShare
https://www.slideshare.net/sylvinus/ranking-the-web-with-spark ▾
Nov 15, 2016 - Ranking the Web with Spark. 1. Ranking the Web with Spark Apache Big Data Europe 2016 sylvain@sylvainzimmer.com @sylvinus; 2.

Ranking the web frontier
citeseerx.ist.psu.edu/viewdoc/summary?doi=10.1.1.387.8325 ▾
by N Eiron - 2004 - Cited by 288 - Related articles
CiteSeerX - Document Details (Isaac Councill, Lee Giles, Pradeep Teregowda): The celebrated PageRank algorithm has proved to be a very effective paradigm ...

The web: fact or fiction, asks Tim Berners-Lee | News | The Guardian
https://www.theguardian.com › News › Tim Berners-Lee ▾
Sep 15, 2008 - The future may be to rank the credibility of websites, says the man who invented the web.

[PDF] Link-Based Ranking of the Web with Source ... - Semantic Scholar
https://pdfs.semanticscholar.org/c9cc/e582fbc61a83670558322ab7d458d401d941.pdf ▾
by J Caverlee - Cited by 4 - Related articles
Ranking the Web frontier. In WWW, 2004. [13] M. Ester, H.-P. Kriegel, and M. Schubert. Accurate and efficient crawling for relevant websites. In VLDB, 2004.

Fig. 4.2 The first page of Google's answer to the query "Ranking the web," August 28, 2017.

Popularity of websites

There are a number of companies that measure and rank the popularity of websites, like Ranking.com, Alexa Internet, comScore, Compete, Quantcast, and Nielsen Holding. Alexa is perhaps the most popular traffic-ranking service today. Alexa's Traffic Ranks are based on the Internet traffic data provided by users over a rolling three-month period, and the list is updated daily. A site's ranking is calculated from two measures: unique visitors and page views. Unique visitors are, ostensibly, determined by the number of unique users who visit a site on a given day. Page views are the total number of user URL requests for a site. Multiple requests for the same URL on the same day by the same user are counted as a single page view. The site with the highest combination of unique visitors and page views is ranked #1.

Nobody is surprised that Google leads the popularity list (does it have a soccer ball?), but there is some rank reversal in second place (Facebook vs. YouTube) depending on the ranking system. A famous example concerns PageRank's method for changing the numerical value of what is called the "damping factor" in order to give different results. PageRank is based on assumptions about how a web-surfer behaves. For a while a web-surfer will click on the links she is seeing on a certain page, but she will get bored with the actual page she visits and then jump to another page randomly (by directly typing in a new URL rather than following a link on the current page). The original algorithm assumed that the probability of being bored is 0.15, so the numerical value of the damping factor was set as 1 – 0.15 = 0.85. So, by setting the damping factor for other numbers we may get different ranking. The phenomenon is called *rank reversal*, which describes a change in the rank ordering depending on some not important, or in many cases irrelevant, factors.

What is hot on the Web?

Reddit is a content-rating and discussion website, self-described by its slogan "The front page of the Internet." As of July 2018, Reddit had 542 million monthly visitors (234 million unique users), making it the third most visited website in the United States and sixth in the world, according to Alexa Internet. Reddit categorizes its content by what it calls hot, new, rising, controversial, top, and gilded. Its algorithm is more or less open source and freely available. Reddit's hot ranking uses the logarithm function (to read this book the only math you should know is the ability to discriminate between algorithm and logarithm—although, for the sake of safety, I should make it explicit that I am joking) in order to weigh the first votes more heavily than the rest. Generally this rule applies: the first 10 up-votes have the same weight as the next 100 up-votes, which have the same weight as the next 1,000, and so on.[34]

Last night (July 2, 2018) soccer fans saw perhaps the most dramatic game in the World Cup competition so far, between Belgium and Japan: the "red devils" were two goals down but won the game by a literal last-minute goal. As I see this morning, the top hot topic on Reddit is "Japanese team leaves a 'thank you' note in Russian inside their locker room despite their heartbreaking 2–3 defeat to Belgium." While soccer fans will preserve memories from the game, still the news itself is expected to be a typical "sensation of the day" and will be overshadowed by other events tonight.

A complex network like the Web certainly contains interesting and important items that are not easy to detect.

About the results of the game: stability, reversal, statistics

Rank preservation and rank reversal

We have seen many examples so far concerning how human cognition and computational algorithms produce ranked lists. Our recurring question is: How reliable and stable are the results? Another is: How should we cope with these lists? And how strongly can we trust a ranking system to find the items of best quality?

Obviously, we can use different criteria for ranking. There is a mathematical procedure called *multicriteria ranking*. It is a complicated process in which not only are the alternatives ranked, but also the criteria themselves must be prioritized. Imagine you are young person who needs a car. The seller shows you two cars: a new one (N1) and a used one (U), and the price of the used one is half of the new one. So, you have to cope with two criteria: price and age. You might be inclined to think that at this stage of your life budget matters much more, and you really need four wheels, so you almost decide to buy the cheaper car U. But while you are just looking around, the dealer shows you another new car (N2). It is somewhat fancier than the other new car, but *much* more expensive. It may happen that you change your decision, thinking, "Well, I'm getting a good deal if I buy N1! I could have spent much more on N2, so I'm basically saving money by buying N1!"

This simple story tells us that it is not true that we don't have to choose in an ideal world. An ideal ranking procedure would preserve ranking between items by adding or deleting new alternatives. This was known as the *principle of invariance* or the *independence from irrelevant alternatives*, and we mentioned this earlier in the chapter in relation to Arrow's impossibility theorem.

The 2000 U.S. presidential election is frequently mentioned as an example of the violation of the principle of independence from irrelevant alternatives. We already know from at least Arrow's impossibility theorem 4 that no voting system is perfect. Al Gore, the Democratic candidate, lost to George W. Bush, the Republican candidate, but he lost in the Electoral College. The decisive vote was in the state of Florida, where the final certified vote count showed Bush with just 537 more votes than Gore. Gore supporters argued that the third candidate, Ralph Nader, spoiled the election for Gore by taking away enough votes from Gore in Florida to let Bush win.

Rank reversal may occur due to the change of importance of the different features. While links and content are the most two important factors that determine a page's ranking, now it is known that Google uses about two hundred other factors. The variants of the PageRank algorithms are based on the collective wisdom of the participants in a network. Network scientists showed that the ranking procedure shows stability against (relatively small) perturbations in a network.[35] It is fun to understand that ranking on the Web is a combination of people's opinion and mathematical algorithms created by humans.

Statistics of ranking

It is a general view that ranking words by their frequencies shows some statistical regularities. This idea is referred to as Zipf's law after the observations of Harvard linguist George Kingsley Zipf (1902–1950), published in 1949, that there exists a proportional relationship among the frequencies of words in texts. The most frequent word will occur twice as often as the second most frequent word, three times as often as the third most frequent word, and so on and so forth. Later it was observed in many languages, that the frequency of the most common words is proportionally related to the inverse of its rank. For example, the word "the" is the most commonly used word in the English language. The second most common, "of," is used about half as much as the first. The third, "and," is used about a third as often as the most common, and so on. The size of cities in the United States (and in many other countries) shows the same statistical pattern, and these are not the only two results: corporation sizes, income rankings, and many other ranking items have similar statistics. It is related to what is called the 80/20 rule: roughly 80 percent of the effects come from 20 percent of the causes. The understanding and management of these ranking patterns is possible and necessary. Similar laws were observed by the Italian economist Vilfredo Pareto around 1900 by studying the distribution of incomes. He noticed that a small proportion of a population owns a large part of the wealth.

These kinds of statistics are very different from the bell curve, which everybody knows. They are called "long tail" or "heavy tail" distributions, and as opposed to the bell curve, which is symmetrical, these distributions are *skewed.* Skewness is a measure of asymmetry of a distribution and describes the deviation from the bell curve. The overwhelming majority of biological, technological, and social networks are characterized by heavy tail distributions. *Preferential attachment* is a term that has been suggested to describe the generation of degree distribution (also called as edge distribution) of evolving networks. It is a simple model with scale-free behavior; the edge distribution follows a *power law.* Such behavior has been found in many networks such as airport networks, scientific collaboration networks, and movie actor networks. The model is very simple, and as the reader certainly knows, it became extremely popular.[36] Scientific citations and artwork prices are also described by power laws, as we will discuss in Chapter 7.

Lessons learned: the scope and limits of our rationality

The term *objective reality* is associated with truth and reliability. Modernism as a philosophy placed trust in the idea that objectivity, truth, rationality, and reliability had a high value. Even for us (moderate) optimists, there is no shame in admitting the limits of our objectivity and of our rationality. Early theories of both individual and social decision making were based on the concepts of rationality and optimization. Results of research in the last 60 years resulted in a shift from the concept of the rational *Homo economicus* to a new model of decision-makers with cognitive biases and a clearer knowledge of their fallacies. Individual choices and preferences are aggregated to form social preferences, and in this chapter we reviewed some techniques behind this aggregation. We also learned that preference ranking does not always imply a unique result because we might get cyclic results, as in the RPS game. It was remarkable to see that elements of this game appeared in both ancient religious systems and in the US governmental system.

Further, ranking algorithms should be the main tools of generating objective rankings, and we all know that Google's PageRank algorithm made a fortune for its inventors. Google became Google because of its algorithm's ability to produce a relevant ranking of websites within a very reasonable time. We understand now that the algorithm could produce different results if modified, and more generally, rank reversal may happen in real-world situations. If we rank many elements based on some characteristic features (e.g., words based on the frequency of their occurrence or cities based on their size), we can use statistical methods. In many real cases, the distribution of these features strongly deviates from the bell curve, and models instead a skew distribution, technically called a power law distribution.

We should understand and accept that whether we are ranking other people, things, or options or are being ranked by others, the result is the consequence of some not totally rational and objective analysis. The next chapter further studies the reasons why measurements and ranking of social institutions might be biased.

5

The ignorant, the manipulative, and the difficulties of measuring society

There are at least two reasons why we may not have objectivity in a ranking procedure. In principle, ranking agents should be objective, but, more often than not, they are ignorant or manipulative. Ignorant agents lack the knowledge of some facts or objects or the skills to do something. They (never *we*!) are not necessarily uninformed; rather, they are *mis*informed.[1] Manipulators change, control, or influence something (or someone) cleverly, skillfully, and generally for their own advantage. The actions of the ignorant and the manipulative construct a deviation from "true ranking," and they give the illusion of reality while producing artificial changes in reality.

The ignorant

Not only cognitive bias

In Chapter 4 we discussed the theory behind different forms of cognitive bias, and here we will discuss real-world illustrations of different types of cognitive biases. Bertrand Russell (1872–1970), a British philosopher, mathematician, and Nobel Prize winner in literature, once said, "One of the painful things about our time is that those who feel certainty are stupid, and those with any imagination and understanding are filled with doubt and indecision."

Much earlier, Confucius (551–479 BCE) stated, "Real knowledge is to know the extent of one's ignorance." The wisdom of these philosophers has since been supported by studies by social psychologists David Dunning and Justin Kruger.[2] They concluded that those with less knowledge suffer from illusory superiority due to their cognitive bias, as Figure 5.1 illustrates.

The Dunning–Kruger effect reflects a very important psychological mechanism underpinning biased ranking. It is well known that competent students underestimate themselves, while incompetent students overestimate themselves, regarding their class standing. Similarly, young drivers grossly overestimate their skills and response times while operating a vehicle. Literary and movie characters often embody the Dunning–Kruger effect, so their ranking ability is biased. Simply put, they cannot correctly estimate their places in their communities. Many experience an unfortunate combination of being uninformed, misinformed, or disinformed (to the memory of Elemér Lábos [1936–2014], a medical doctor

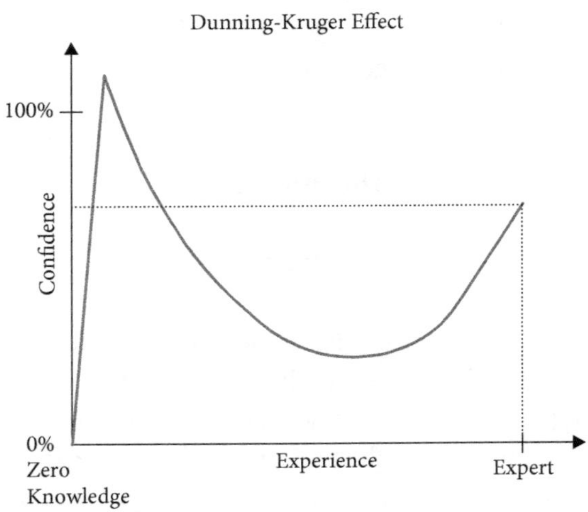

Fig. 5.1 The non-monotonic relationship between self-confidence and expertise. (Image via Wikimedia Commons.)

and mathematician).[3] Probably the worst-case scenario in regards to ignorance entails having misleading mental models composed of false theories, facts, metaphors, intuitions, and strategies that one might regard as useful knowledge. (I cannot resist referring to "The Dunning–Kruger Song" from *The Incompetence Opera*.[4] The video, linked in the notes, is three minutes long, and it is worth watching.)

One movie character who embodies the Dunning–Kruger effect is Rodney Farva from *Super Troopers*. He is a rather terrible cop, but he gets really excited to be involved in whatever the team is doing and insists on "helping out," while it is obvious to everybody else that he's not really helping. (See "Best of Farva" on YouTube for context.[5])

While I am far from praising ignorance, I can acknowledge that it might have benefits or lead to success. Christopher Columbus is known to have gone searching for a new path to Asia and instead discovered a new continent. A young Swedish guy, Ingvar Kamprad, who owned a mail-order company, once tried to fit a table into his car in order to sell it, and on the suggestion that he remove the legs in order to transport it more easily, he got the idea for flat-packed furniture, which led to the emergence of IKEA. Totally new companies, like Amazon, Uber, and Airbnb have revolutionized industries by ignoring the traditional knowledge of well-established companies in the bookselling, taxi, and hotel industries. So we can see that some ignorance combined with striking new insight has the potential to introduce innovative ideas. But what happens when the ignorance is too much?

"When the world is led by a child"

If you Google "Dunning–Kruger president," you will find several thousand websites in your search results.[6] In a *New York Times* op-ed article titled "When the world is led by a child," David Brooks writes,[7]

> He is thus the all-time record-holder of the Dunning–Kruger effect, the phenomenon in which the incompetent person is too incompetent to understand his own incompetence. Trump thought he'd be celebrated for firing James Comey. He thought his press coverage would grow wildly positive once he won the nomination. He is perpetually surprised because reality does not comport with his fantasies.

Indeed, Michael Wolff's provocative book about the White House[8] during the Trump presidency is not genuinely surprising. It is about the president's intellectual limitations, his horrifying ego, and his immature need for constant recognition and verification. What is the probability that the ignorance of the president will have any positive effect? I don't know; I am not an expert.

The manipulative

We have seen huge scientific advances from quantum computing to space exploration during the 20th and 21st centuries. It would be silly to assume that psychology did not develop concurrently with these other sciences, giving us a greater understanding of the psychology of manipulation.[9] If we consider the ranking game a competition, some players are ready to violate the rules to ensure their advantage or priority. If the rules are unwritten, it is even easier to breach the regulations. In many games there are referees, umpires, judges, or arbitrators. However, "life is a game with many rules but no referee," as we know from Joseph Brodsky (1940–1996), the Russian-American Nobel Prize–winning poet. (More precisely, he is well known for his answer to the question: "You are an American citizen who is receiving the prize for Russian-language poetry. Who are you, an American or a Russian?" to which he responded, "I'm Jewish, a Russian poet, an English essayist—and, of course, an American citizen.") Manipulators have the intention of gaining personal advantage by adopting different tricks, from outright cheating to sophisticated propaganda techniques. But in all cases their goal is to reach the top of the "success list" once again, even by violating the rules.

How to manipulate

Appeal to fear

The appeal to fear is a technique used to motivate people to take a specific action or support a particular policy decision by arousing fear. The reader knows well that this strategy has been adopted by more than a few US presidents with appeals like, "If we don't bail out the big automakers, the US economy will collapse. Therefore, we need to bail out the automakers." Experts said the argument was an exaggeration, but it worked nonetheless. Others have made appeals such as, "The attacks on our police, and the terrorism in our cities, threaten our very way of life." President Trump's "formula is very clean and uncomplicated: Be very, very afraid. And I am the cure."[10]

Similarly, the 2018 Hungarian election campaign had a single topic: fear. According to political analysts in Hungary, "It hardly seems to matter that the migration crisis has largely passed and that there are now more posters in Hungary about the danger of immigrants and refugees than actual refugees and immigrants let into the country this past year. The poster is in keeping with a campaign that has been rife with dirty tricks, false news stories, vicious personal attacks, conspiracy theories and perceived enemies all around."[11]

Black-or-white fallacy

Another US president declared in reference to the "war on terror," "Either you are with us or you are with the terrorists."[12] If asked to choose between opposing the Patriot Act and being a patriot, the question necessarily implies that if you are against the Patriot Act, then you cannot be a patriot. However, these black-or-white fallacies ignore the nuance involved in such questions, and they fail to recognize that if someone is not your ally, it is not necessarily the case that she is your enemy. You cannot exclude the possibility that this person might be neutral or simply undecided. If you are forced to choose between two options, to the exclusion of all other possibilities, this presents a logical fallacy.

Selective truth

Selective facts are more dangerous than fake news. We use the news to make decisions by ranking (consciously or unconsciously) the options before us, based on what is occurring in the world around us. The media mogul Rupert Murdoch once declared his purpose: "Produce better papers. Papers that people want to read. Stop having people write articles to win Pulitzer prizes. Give people what they want to read and make it interesting."[13] As was discussed in the previous chapter, we are subject to confirmation bias, so we prefer to read news that fits into our preexisting mental frameworks. While news has traditionally been intended to accurately reflect the state of affairs in the world, news filtering mechanisms that amplify our existing beliefs and biases have become increasingly popular. To put it another way, media organizations try to maximize their readerships and viewerships by finding out (using data and algorithms for efficiency) what kind of news we engage with most frequently and by reproducing and feeding us this kind of news.[14]

How do we react if we get selective facts from the other side? Actually, I have some fresh experience with this phenomenon. I am writing this section in July 2018 in Budapest, Hungary. As you already know, I am a soccer fan, so I am watching almost all games of the World Cup on the state-controlled Hungarian sports channel. During halftime there are short news segments, which all repeat stories about crimes committed by immigrants somewhere in Europe. The Hungarian leader has learned the next lesson: repetition, repetition, and repetition (of oversimplified and one-sided) messages!

Repetition

Lewis Carroll (real name: Charles Lutwidge Dodgson [1832–1898]) wrote in *The Hunting of the Snark*, a nonsensical poem: "What I tell you three times is true." While English literature was probably not his strength, Hitler famously once claimed that there is "no limit on what can be done by propaganda; people will believe anything, provided they are told it often enough and emphatically enough, and that contradicters are either silenced or smothered in calumny."[15] We are familiar with the political slogans of our time, from "Yes We Can" to "America First." In Orwell's *Animal Farm*,[16] Old Major repeats the same idea with slight stylistic variation to argue against the humans: "Man is the only real enemy we have"; "Remove Man from the scene and the root cause . . . is abolished"; "Man is the only creature that consumes without producing"; "Only get rid of Man." More systematic psychological studies show that repetition creates the "illusion of truth."[17] My own suggestion is this: please don't repeat things without carefully checking whether or not they are true. If you do, you are also responsible for creating a world where it is difficult to discriminate between lies and truth. So, please, please, please, think before you repeat!

Appeal to authority

We cannot say that it is unreasonable to believe authorities. There is a logical model behind the appeal to authority:

- Assumption 1: X is an authority on a particular topic.
- Assumption 2: X makes a statement about that topic.
- Conclusion: X's statement is probably correct.

In the ideal world of scientists, there is an agreement that authorities should prove their statements as rigorously as a graduate student would. In politics, however, this is not necessarily the case, and everybody is familiar with this fallacy: "an extremely credible source" has called my office and told me that Barack Obama's birth certificate is fraudulent. Of course the statement is extremely credible, since the source stated so himself. How about this statement? Einstein said that $E = mc^2$, so it is true. There is no causal relationship between who says something and whether or not it's true. What is true is that mass–energy equivalence is a general principle, and it is the consequence of some fundamental properties of time and space. But no more physics; let us speak about ads. What is the relationship between a celebrity actor like Dr. Ross and machines that brew coffee? Celebrity testimony is often used to heighten the appeal of a particular product, as celebrities are considered experts when it comes to fine products. Dr. Ross claims that "he is proud to work with [Nespresso] in its commitment—that every cup of its coffee has a positive impact on the world."[18]

Game change in media manipulation

Mark Twain once said, "If you don't read the newspaper, you're uninformed. If you read the newspaper, you're misinformed." Even in the 19th and 20th centuries, Mark Twain's era, there were threats to media objectivity. Politicians and journalists might have wanted to change reality, and they could exploit the fact that the media was more or less reliable. However, at that time, distortion, exaggeration, fabrication, and simplification were the exception, not the rule, as they are now.

Traditional authoritarians controlled all media, and they adopted censorship and ideologically oriented propaganda to maintain hegemony over their populations. In the world of the new authoritarians, more sophisticated methods are employed to influence public opinion and shape political narratives. By restricting space for alternative media outlets, and ensuring the dominance of state-owned and state-friendly media assets, the new autocrats keep dissenting views out of the news and manipulate political discourse.

In the past, general-interest intermediaries (think of the 1960s, when there were three major news networks—ABC, NBC, and CBS—that controlled TV news), to use Cass Sunstein's term for such media giants,[19] have exercised a great deal of influence over access to information, and, as Zeynep Tufekci points out in a stunning piece for *Wired* magazine in 2018, traditional techniques of censorship have entailed shuttering newspapers, revoking broadcast licenses, or threatening (or even murdering) journalists who disagreed with a government's agenda. But now we live in an age where "media" has come to mean everything from CNN or NPR to one's Facebook feed. With the decline in the relative power of general-interest intermediaries and the rise in the influence of personalized media delivered by Google News, Apple News, and a host of other curators, manipulation and censorship techniques now focus on making the entire media landscape seem illegitimate and sowing distrust in what have historically been considered "objective" institutions and voices.[20] Despite technological developments that have allowed this sort of manipulation to proceed with a degree of ease never seen before, the phenomenon itself is not entirely new. In her 1967 essay "Truth and politics," published in the *New Yorker*, Hannah Arendt pointed out that the function of repeated lies in (specifically, political) discourse, which are increasingly common today, is to cast doubt on the very reality we inhabit:

> In other words, the result of a consistent and total substitution of lies for factual truth is not that the lies will now be accepted as truth, and the truth defamed as lies, but that the sense by which we take our bearings in the real world——and the category of truth vs. falsehood is among the mental means to this end——is being destroyed.[21]

The future of free speech and new forms of censorship and manipulation are now hot topics in our changing world,[22,23] as a new type of manipulation has seemed to emerge. We read what we want to read, and it is a major threat to our political institutions.[24] Since technology allows for the efficient use of filtering, people receive predetermined information, delivered in a personalized journal, the "Daily Me." So, they (we) live in echo chambers, which are means for amplifying our beliefs. (Remember the impact of confirmation bias!) These echo chambers exclude the possibility of receiving surprising news from people living in other chambers, and they contribute to life in a society where people's minds are closed, by their own choices, to other opinions and beliefs.

Manipulation in movies, history, and elections

The word "manipulative" has a natural negative connotation to it. To manipulate does not just mean to convince; it implies an influence by deception, to unfairly control a person by exploiting him. Therefore, when we think of the most manipulative movie characters in cinema history, we immediately think of villains. A good guy doesn't manipulate; he persuades or influences. But a bad guy deceives, lies, and schemes. *Ranker* gives first place to Keyser Söze from the 1995 film *The Usual Suspects* in the category of "the most manipulative characters in film." Söze is a ruthless and influential crime lord who acquired a legendary status among both police and criminals.[25]

I am not sure how to compare historical figures with regard to how manipulative each is. Machiavellian characters throughout history have used tactics along the lines of "the ends justify the means." To put it another way, legal and moral rules can be violated in order to reach a very important, sufficiently justified final goal. Quora gives the following list of manipulative figures in history, but it is probably best to consider it an unordered list:[26]

- Adolf Hitler
- Joseph Goebbels
- Charles Maurice Talleyrand
- Otto von Bismarck

- Albert Speer
- Henry Kissinger
- Joseph Stalin

There are many countries one could speak of when discussing election manipulation, but for brevity's sake I will focus on Zimbabwe.[27] Researchers have studied the 2013 presidential elections in Zimbabwe, in which Robert Mugabe, the country's long-time leader, won a seventh presidential term with more than 60 percent of the vote, a startlingly large margin considering the large discrepancies between survey predictions and official results. Analysis has generated the following conclusions:

- Most likely, the incumbent's margin of victory in Zimbabwe in 2013 was far smaller than reported.
- Much electoral manipulation occurred before elections.
- Much electoral manipulation occurred in rural locations.
- Fearful voters were intimidated to support the incumbent president's party.
- The aggregate extent of manipulation at the national level accounted for one-sixth to one-fifth (from 16 to 20 percent) of the total reported vote.

Like it or not, ignorance and manipulation are omnipresent in human society. When we try to understand with our limited minds the complexities of human society, we often turn to measurement and quantification as a useful heuristic. Again, we will see the difficulties of being objective.

The importance and the difficulties of measuring society

We will turn here to more formal observations and laws about the illusion of objectivity. In the US, such observations have been formulated as Campbell's law, while in the United Kingdom, they are referred to as Goodhart's law; however, the two refer to essentially the same principle.

The reality and myth of measurement

The process of measurement was indispensable even in ancient civilizations. The determination of quantities such as length, mass, volume, and time was crucial for supporting agriculture, construction, and trade. William Thomson (1824–1907), generally referred to as Lord Kelvin, famously stated: "When you can measure what you are speaking about, and express it in numbers, you know something about it. When you cannot express it in numbers, your knowledge is of a meager and unsatisfactory kind; it may be the beginning of knowledge, but you have scarcely in your thoughts advanced to the stage of science." Frederick Taylor (1856–1915) founded what is known as *scientific management* and adopted the practice of measuring production-related labor processes with the hope of improving productivity. This approach, called *Taylorism*, was attacked for the perception that it considers workers to be "cogs" in the big machine of the factory and was famously mocked in Aldous Huxley's *Brave New World* (1932) and in Charlie Chaplin's *Modern Times* (1936). However, its spirit survived, and the belief persists that "Measurement is the first step that leads to control and eventually to improvement. If you can't measure something, you can't understand it. If you can't understand it, you can't control it. If you can't control it, you can't improve it."[28]

The dangerous side of measurements

Donald Campbell (1916–1996) was a social scientist with an extremely broad field of interests. Campbell's law,[29] as it is commonly called, states, "The more any quantitative social indicator is used for social decision making, the more subject it will be to corruption pressures and the more apt it will be to distort and corrupt the social processes it is intended to monitor." Similarly, Charles Goodhart, an economist from the London School of Economics and a former member of the Bank of England's Monetary Policy Committee, stated that "once a social or economic indicator or other surrogate measure is made a target for the purpose of conducting social or economic policy, then it will lose the information content that would qualify it to play that role." Goodhart's law[30] holds that "any observed statistical regularity will tend to collapse once pressure is placed upon it for control purposes."

Managers in every area, from law enforcement to health care, travel to education, have to report numbers to characterize the performances of their organizations. There are many well-documented examples from the former Soviet Union and related countries that could be used as case studies for Campbell's law. Economic planners set targets for their factories, emphasizing quantity rather than quality, and directors were judged on whether or not they hit their quantitative targets. Product quality and consumer satisfaction were not major factors, and, as a result, "When five-year plans set targets in terms of tonnage, factories made things that were comically heavy—chandeliers that pulled down ceilings and roofing metal that collapsed buildings."[31]

Tyranny of metrics?

I was already working on this project when the closest book to my subject was published.[32] Jerry Z. Muller's *The Tyranny of Metrics* studies our obsession with metrics, and Muller lists what he calls the unintended consequences of such an obsession. Muller might be correct in his argument that a lack of social trust is the main reason that human judgments have been substituted by metrics describing accountability and transparency. In a world where we assume the honesty, integrity, and reliability of other people, transparency could be reached by using fewer metrics.

Everybody knows stories about how metrics have been gamed. In policing, the number of cases solved, crime rates, and other statistics have been manipulated to produce a better image of the performance of a police department. In education, "teaching to the test" works against the real goal of schooling (education) in order to meet externally prescribed targets. In the health care system, we have heard anecdotes about surgeons who avoid treating risky patients so as not to reduce their performance measures. Muller also is right that there is a discrepancy between what can be measured and what is worth measuring. One example is that it is easier to measure the amount of investment than it is to measure the result of an investment.

While I do agree with the overwhelming majority of the examples and arguments in his book, as an ardent scientist, I can't comply with the tone of his conclusions. Would it be a welcome development to abandon the use of metrics, rating, ranking, and any quantitative analysis? Who would then make judgments, and what would be the basis of such judgments? I think the book neglected to analyze the benefits of the accountability provided by metrics, which might overcompensate for the obvious drawbacks. Well, Professor Muller, I am ready to offer a draw.

Observers and observed

The reality of science is based on objective measurements: experiments with results that are reproducible. In science, it is rare for somebody to state, and receive respect for, self-congratulatory declarations, and there has probably never been a scientist who declared of himself or herself, "I am such a fantastic stable genius that I am able to make an experiment that nobody can reproduce!"

In the overwhelming majority of cases in natural science, the observed phenomenon (say, the velocity of the fall of an apple from a tree) does not depend on the mental state of the observer. Even if you sleep, the apple will fall down. (And I hope the apple did not fall on your head.) In the world of microscopic particles there is an interaction between observer and observed (but I am not writing a book on physics).

Unfortunately, humans are not apples. Observations influence human behavior. Even infants may be more prone to crying if they know it will get them what they want. Campbell's law and Goodhart's law are nothing more than illustrations of a famous quote attributed to the physicist Murray Gell-Mann (1929–2019) (who received a Nobel Prize for his contribution to theory of particles): "Think how hard physics would be if particles could think!"

The bottom line is a triviality. Observations, measurements, and assessments reflect the past performance of people and institutions. However, people and institutions have the chance to act and react. They adopt strategies for generating a better-than-real result by manipulating information (say, if police don't report all the crimes). However, the goal of the majority of performance assessments is to help decision-makers allocate resources. The most frequently used resource allocation strategy is to give funds to those competitors who showed a better performance, giving them a better ranking. This reactive mechanism leads to the amplification of small advantages.

Matthew effect, positive feedback, reinforcement

The sociologist Robert Merton coined the term "Matthew effect" to refer to the mechanism by which small social differences are amplified.[33] Paraphrasing the Gospel of St. Matthew 25: 29,[34] Merton sought to explain why and how well-established scientists are able to dramatically increase their resources in comparison with those who are less well established. The core of this mechanism is called *positive feedback*, which implies reinforcement, and it often leads to unintended consequences. Specifically, it works against competition, according to some social researchers.[35] The allocation of human and material resources to people and institutions, which results from the competition for resources in the wake of the Matthew effect, leads to an ever-growing inequality and a resulting restriction in competition. The "losers" become too poorly equipped to challenge the winners, and if the majority of the competitors die out, an oligopolistic competition with few rivals will survive.

A large family of unintended consequences is related to the *cobra effect* or *rat effect*. The rat effect takes its name from an unfortunate pest situation in Vietnam many years ago. When the French colonized Hanoi, Vietnam, the city was full of rats. To reduce the unbearable concentration of rats, a policy was instituted in which people were paid to kill the rats. People had to present only the tail of the animal to receive payment for their services, and a new strategy emerged where people did not kill the rats; they merely chopped their tails off and then let them back into the sewers, allowing the rats to continue breeding and ensuring more profit to the rat killers.

Similarly, the term "cobra effect" was born during the British colonization of India. To fight against the large number of poisonous snakes, the British administration offered a premium for every snake killed. The policy worked well initially, but local people started breeding the snakes to ensure their continued ability to receive the premium. The administration stopped the program, but it was too late, for the region was filled with even more cobras than before. Generally, the cobra effect describes situations where a proposed solution to the problem makes the problem even worse.

Another well-known example concerns cities that wish to ban vehicle traffic. Mexico City and Bogotá have introduced policy measures in an attempt to reduce traffic that allow car owners to operate their vehicles only on specific days of the work. Bogotá has been working to eliminate cars on city streets since 1974. For example, on Monday, no cars were allowed on city streets with license plate numbers ending in one through five. As a result, those who felt that they absolutely must drive each day bought a second car so as to avoid the restriction. Not surprisingly, these were old cars, and the unintended consequence was more congestion and, accordingly, more pollution. Still, the policy has been very popular with voters, and recent news has reported that 13 cities (including Madrid, Oslo, London, and Brussels) are instituting policies with the intent of banning cars.[36] We'll have to wait a few years to see the results, but we can make an educated guess as to what they will be.

Social metrics: useful, but not a silver bullet

Campbell, Goodhart, and others don't say that numbers and quantitative evaluations are bad. What they *do* say is that numerical evaluations not only reflect the past but also influence the future. The fear of receiving a low ranking affects a manager's decisions about her future actions. We cannot deny that numerical data are susceptible to manipulation or distortion. However, it does not mean that we should give up when it comes to using datasets to improve social programs and institutions. It seems to be true that the increase in high-stakes testing leads naturally to an increase in cheating by test-takers, but I don't believe this constitutes a sufficient reason to replace these tests with more subjective methods of evaluating student progress. However, we should think more carefully when it comes to using quantitative data in decision making.

So what's the alternative? Well, in my humble opinion it's about treating measures as one would treat a barometer and better understanding causality. If your barometer tells you it's going to be low pressure, you'd be advised to take an umbrella. On the other hand, you'd be crazy to wear shorts and a sunhat in an attempt to raise the atmospheric pressure, and taking the barometer and placing it deep underwater to raise the recorded pressure is stupid too.[37]

Rank and yank

"Rank and yank" refers to an annual performance review process by which a company ranks its employees against one another and subsequently uses these rankings to make life-changing decisions for employees. Often, the firm terminates the employment of the people at the lowest end of the ranking. Is ranking and yanking so brutal an activity? Well, even in elementary school, children are graded on their performance, and their place in the formal class hierarchy is precisely determined, but perhaps the consequences are not so severe.

Jack Welch, the former CEO of General Electric (founded by Thomas Edison) employed a rigid system of rank and yank in which the bottom 10 percent of employees were fired. Managers were simply forced to decide where their employees fell in the hierarchy. We would be lying to deny that the forced-ranking system worked, at least for a while, because it helped employees know where they stood in the corporate hierarchy. Generational changes and technological development both required and made possible the transition to new evaluation systems.[38] I am ready to accept the belief that the current generation needs feedback more frequently than on an annual basis, and GE is now implementing a system of daily feedback via an app (PD@GE).[39] The hope is that the former system of "command and control" management, which implies too much competition among employees, will be replaced by a system that increases cooperation among them. Let us leave it to the future to find the healthy balance between competition and cooperation.

The infamous former CEO of Enron, Jeffrey Skilling, was motivated by Richard Dawkins's book *The Selfish Gene*, a controversial and very influential text on evolution,[40] which states that the unit of natural selection is the gene and not the organism. Skilling's managerial philosophy was driven by his belief that money and fear are the only means to motivate people. In Enron's performance review system, employees were annually graded from one to five, with five being lowest. There was a relative, comparative element in this system: 15 percent of people had to be graded five, regardless of absolute performance. (I am a college professor in the age of grade inflation, so, the "cynical me" raised a question to myself: Would it be difficult to hand out failing grades to 15 percent of my classes if my provost instructed me to do so?) The review process was considered the most important element in the life of the company. Dawkins, in response, made clear that Skilling misunderstood his book, and he has never suggested selfishness is the driving force of progression.

While Marissa Mayer served as the CEO of Yahoo!, she introduced the quarterly performance review. During the time of the #MeToo movement it is remarkable to read about gender discrimination against males, but some male former employees sued the company, alleging discrimination against men in the opaque review process.[41] The lawsuits were ultimately dismissed. Of course, there is no causal relationship, but in any case, Yahoo! is not an independent company anymore, and it is owned by Verizon.

It looks as though forced ranking systems encourage competition among employees, but CEOs still have the difficult problem of avoiding unhealthy *dog-eat-dog* situations in the workplace. But now we will turn to another social metric that strongly influences all of our lives, not just those of us in employment situations with forced ranking: credit scores.

Credit scores

A little history

Buy, buy, buy! We want to buy things even when we cannot afford them, so we ask somebody to lend us money for our purchases. This "somebody" (whoever she is, our friend or a bank) has a single question: "Can I trust that the borrower will repay their loan?" In a world when people did not live in an ocean of data, potential lenders characterized *qualitatively* their potential borrowers: "He looks like a nice, reliable guy, so I think he will repay. In addition, he promised to pay the original sum and *x* percent of interest." Owners of corner grocery stores developed skills over the course of centuries to classify their clients as reliable or not reliable. I find it interesting but not surprising that the oldest credit-reporting agency in the United States emerged from the grocery business.

Cator Woolford was a grocer in Chattanooga, Tennessee. He collected data from his years of interactions with customers, produced a book, and sold copies of the book to the local Retail Grocers' Association. Based on his success, together with his younger brother Guy, a lawyer, the Woolford brothers opened a very small business in Atlanta, which they called the "Retail Credit Company." This small business steadily grew into what we now know as Equifax, Inc., one of the three giant consumer credit bureaus (the other two are Experian and TransUnion), which collects and process information regarding over 800 million individual consumers.[42]

When people are given the task of judging other people's character, it is truly, truly subjective. Granting or denying loans or credit requests was very far from being objective, and age-, gender-, or race-based discrimination happened again and again. To help the decision-makers by instituting quantitative analysis was a big step toward objectivity. The goal has been to eliminate subjectivity, including subjectivity attributable to cognitive bias. William R. Fair (1923–1996) and Earl Isaac (1921–1983) were the pioneers of building mathematical models for predicting the behavior of the potential borrowers, and an initial version of a credit application scoring algorithm was introduced in 1958. This algorithm classified three possible behaviors attributable to borrowers: the borrower will pay on time, will pay with delay, or will not pay at all. The Fair Isaac Corporation was later established and developed an algorithm and related software to calculate what became the famous/infamous FICO score.

How is a credit score calculated, and how objective is the result?

The main goal of this book is to uncover the hidden rules behind our navigation between subjectivity and objectivity. We cannot deny (and I don't have any intention of doing so) that algorithms are based on human assumptions. After these assumptions are made, the evaluation is the outcome of an automatic procedure. To generate a credit score algorithm, the first question is to decide which *input data* should be taken into account. FICO uses five factors:

- The history of how you paid your bills
- How much money you owe on credit cards, mortgages, loans, etc.
- The length of your credit history (the longer the better)
- The mix of your credit (the more diverse the better)
- New credit applications (don't open too many new accounts too fast).

The next (and natural) question is whether or not it is reasonable to assume that all the five factors have the same importance. Assuming the answer is "yes," we assign to each input variable a 20 percent weight. However, it is more plausible to assume that there are more and less important factors, and FICO uses the following weights:[43]

- Payment history: 35 percent
- Amounts owed: 30 percent
- Length of credit history: 15 percent
- Credit mix in use: 10 percent
- New credit: 10 percent

We already know the factors that the calculations take into account, but it is equally as crucial to know which factors *don't* count. The Equal Credit Opportunity Act (ECA), prohibits creditors in the United States from discriminating based on race, color, religion, national origin, sex, marital status, and age. Credit scores are applied in a number of other countries, with a similar goal (i.e., represent the creditworthiness of individuals), but the legal environment might vary from country to country. There are variations regarding how to calculate the credit score. Somewhat more technically speaking, the credit score (i.e., a single number) is the output of the algorithm, and the simplest way to get such a score is by summing the weighted inputs. FICO uses a scale from 300 to 850, but the company is not totally transparent with how the score is calculated. As one blogger writes: "FICO should disclose what goes into its all-important algorithms. They say they don't want people to game them, but considering their importance in buying a house or a car, it can't be a black box that only FICO knows."

The hidden rules of the credit score game

Credit scores have come under scrutiny, like many algorithmic procedures, for concealing, rather than eliminating, certain forms of bias. Experts at one of the world's leading law firms, White & Case, explained in their paper "Algorithms and bias: what lenders need to know" that even clearly unintentional algorithms directing financial technologies may lead to discriminatory decisions. Why? Creditors and lenders have access now, in the age of Big Data, to so-called nontraditional data, such as Internet activity, shopping patterns, and other data that are not necessarily directly related to creditworthiness. Often, these data are analyzed using recently popularized machine learning techniques.

Traditional algorithms use rules of arithmetic and logic, defined by the designer of the algorithm. For example, IF Borrower payed back her previous credit without any delay THEN increase her credit score by *x* points. But machine learning techniques do not rely on a previously defined algorithm. Instead, they *generate* algorithms based on patterns found in large datasets. Take, for example, the approval of a loan request. The software has stored and analyzed data related to the financial behavior of many thousands of previous customers. Loading the credit history of a new applicant as input data, a machine learning algorithm might calculate the output, something like the probability that the applicant will default, based on the patterns it has identified in the previous collection of data.

There are justified concerns that algorithms might make biased decisions, especially those that are unfavorable for already-marginalized groups. Ideally, the decision-makers should take into account only the data permitted by ECA when evaluating a borrower. But we live in networks surrounded by our neighbors, friends, and peers. So if creditors can analyze the data of your social network friends in their evaluation of your creditworthiness, it could inadvertently lead to discrimination based on the data creditors are not permitted to consider. Home addresses might be significant factors, and one's ZIP code is considered a dangerous variable due to longstanding policies like *redlining*, which refers to discriminatory housing practices that historically segregated neighborhoods in many American cities.

A machine learning algorithm may find that there is a correlation between your creditworthiness and the financial behavior of your friends or neighbors. It is a complicated situation: a creditor cannot deny your request on the basis that many of your friends were late in repaying their loan. Further, they should be able to explain the basis of their decision to deny your credit request. But if nontraditional data are used, it is very difficult to give transparent and understandable explanations.

It should be clear that data from your social network cannot be used for evaluating your financial future. However, we choose many of our future activities based on recommendation systems. These recommendations influence our choices of hotels, restaurants, dating partners, and movies, just to give an unranked list. Recommendation algorithms use data regarding "stuff that *my friends like*" to generate predictions for your tastes. We will discuss these algorithms in more detail in Chapter 8.

Should we like algorithms? If you are ready to answer the question with "no," would we be better off by returning to the personality-based, totally subjective credit evaluations?

Developments in financial technologies continue to push the boundaries of what has traditionally been acceptable in issuing loans. In May 2016, the Obama administration's Treasury Department issued a white paper titled "Opportunities and Challenges in Online Marketplace Lending." In addition to the traditional players, online marketplace lending companies have emerged to offer faster credit for consumers and small businesses. It was good news that the Treasury Department found it important to analyze the opportunities and risks presented of this new type of credit system.

Toward fair algorithms?

Computer scientists have realized that algorithms might lead even unintentionally to discrimination. Data-mining methods are based on assumptions that come from the pioneers of modern science, such as Galileo, Kepler, and Newton, all of whom heralded the ability to look to the past to predict the future. While this method worked wonderfully to predict the motion of the planets, should we also assume that historical data regarding social behaviors are useful bases for prediction?

There are now algorithms that forecast crimes based on historical data. Patterns related to the time of the day, seasonality, weather, location (vicinity of bars, bus stops, etc.), crime level in the past, and similar data help police departments distribute their resources for preventing potential crimes. As always, while the goal of *predictive policing* promises to be race-neutral and objective, there are also justified concerns that the application of the algorithmic approach leads to the emergence of new problems related to security, privacy, and the constitutional rights of citizens.[44] Again: algorithms behind predictive policing—much more often than not—help allocate law enforcement resources, but they are not silver bullets to eliminate crimes and can have unintended consequences. As a Lithuanian data scientist named Indrė Žliobaitė, now at the University of Helsinki in Finland, writes in a position paper about "Fairness-aware machine learning":[45] "Usually predictive models are optimized for performing well in the majority of the cases, not taking into account who is affected the worst by the remaining inaccuracies." Since we all know that there are human faces and fates behind the numbers, this poses difficult questions and serious concerns.

We know the horror stories, and I decided not to repeat them, when intentionally neutral algorithms produced sexist or racist output. One reason is that machines learn by examples extracted from the data, and data generated by humans reflect human biases. So it may happen that algorithms may sustain, or even amplify, prejudice and social hierarchy.

Social scientists and computer scientists should cooperate to generate "ethical algorithms." Ethics (moral philosophy), as a discipline, investigates what constitutes "good" or "bad" behavior. (I leave the answers for the philosophers.) From the perspective of machine learning, the question is how to train algorithms to make moral decisions so that data can be preprocessed and unethical data eliminated. We may expect many future studies to be conducted in order to understand the scope and limits of building ethical algorithms, and we should accept that "fairness" is far from being a well-defined concept.

Beyond the algorithms: the lending circles and the credit score game

Algorithms are not the only ones who can learn; people can as well. A researcher at the University of Arizona, Mark Kear, describes and analyzes an example related to how immigrants learn that (1) they should play the credit score game and (2) it is possible to improve their credit history.[46] Kear was a participant and observer of a lending circle. Lending circles are organized by the Mission Asset Fund (MAF), a San Francisco–based nonprofit organization dedicated to helping increase the credit scores of low-income families. In these groups, participants learn strategies for reporting data that will improve their creditworthiness. MAF's techniques have managed to increase participants' credit scores significantly (with a 168-point increase in one case study).

Instead of summary

As John von Neumann wrote in his paper "Can we survive technology?":[47]

> All experience shows that even smaller technological changes than those now in the cards profoundly transform political and social relationships. Experience also shows that these transformations are not a priori predictable and that most contemporary "first guesses" concerning them are wrong. For all

these reasons, one should take neither present difficulties nor presently proposed reforms too seriously. . . .

The one solid fact is that the difficulties are due to an evolution that, while useful and constructive, is also dangerous. Can we produce the required adjustments with the necessary speed? The most hopeful answer is that the human species has been subjected to similar tests before and seems to have a congenital ability to come through, after varying amounts of trouble. To ask in advance for a complete recipe would be unreasonable. We can specify only the human qualities required: patience, flexibility, intelligence.

Lessons learned: Why is it so difficult (but not hopeless) to measure society?

Making objective rankings sounds like an appealing goal. However, as we saw in this chapter, there are at least two reasons why we may not have objectivity: ignorance and manipulation. As we see nowadays often, incompetent people overestimate themselves because of the phenomenon known in social psychology as the Dunning–Kruger effect.

Omnipresent in society is not only ignorance but also manipulation. The process of measurement has a major role in any civilization. According to the optimistic perspective of positivism, measurement is the first step in making improvements. The social demand for accountability and transparency has made quantitative metrics a major tool for characterizing the performances of social institutions. However, Campbell's law is a warning signal that metrics can be (and often are) gamed. Algorithms underpinning "rank and yank" and credit scores are illustrative examples of the fact that such mechanisms can amplify social inequalities. However, we should not abandon algorithms in favor of our previous subjective and verbal evaluations. Instead, social scientists and computer scientists should cooperate to generate "ethical algorithms."

Now the reader is familiar with the possibilities and difficulties of the ranking game, which we all play. The next chapter discusses two major instances of the game: university rankings and country rankings.

6

Ranking games

The top-10 illusion

The magic power of round numbers or left-digit effects

We discussed in Chapters 4 and 5 the sources of cognitive bias and its effect on our thoughts and behaviors. Because we are presented with an abundance of information and limited capacity for making decisions, we use mental shortcuts or *heuristics* to help us act on the information presented by the world around us. One such heuristic that we previously mentioned is satisficing, seeking a "good-enough" decision rather than a perfect one. But there are many heuristics we unconsciously use in our day-to-day lives as a means of compensating for our cognitive shortcomings. This points to our fascination with lists.

As we already know, we process many (generally long) lists of items each day. Brands are often ranked in lists published by organizations like Fortune 500, *Travel & Leisure*, or ESPN, and these rankings can affect consumer decisions. We also know that our brain comprehends the incoming information so that we may make decisions about which products to buy or which teams to root for. From our perspective, it is important to realize that our brain generates subjective categories. Ranked lists, like those presented by *Vogue*, *GQ*, and countless other magazines and media sources, already contain organized information, but internally, we further categorize the information once we receive it. The science of marketing psychology has evolved to study precisely how consumers subjectively categorize ranked lists and subsequently inform corporate marketing decisions with the information.[1] Much scientific research has supported the conclusion that *round numbers* are extremely important to the manner in which we perceive numerical information. When we see that an item is ranked 10th, we feel that it is closer to the 8th than to the 11th item. This is just one way in which marketing strategists exploit our bias toward round numbers, and it demonstrates how our perception can be manipulated.

As another example, in the majority of Western cultures, we process numerical information from left to right. Thus, $19.99 is seen as meaningfully less than $20.00 because that leftmost "1" is coded by our rapid decision-making functions as smaller than the leftmost "2." After we make a rapid decision, the slower, more analytical part of our brain recognizes that the difference of a cent means nothing, but it's too late—we have already fallen victim (again) to cognitive bias.

Imprecise information can be more efficient than precise information

Our naïve belief in rationality leads us to assume that the more precise information we have, the more easily we can obtain an objective image of reality. Marketers, however, have long taken advantage of the fact that this is not necessarily the case. Of course, marketers' goal is to cast their brand in its most favorable light, and they found that it was more effective to be a member of a "top-10 tier" than to be explicit about a brand's rank as ninth in its category. Therefore, instead of referring to the exact rank of a brand, they often simply communicate the brand's membership in a general tier along with other top brands. For example, studies have shown that a large number of MBA programs (if you wish to have more precise information, 72 percent of them) intentionally publish imprecise information about their standing compared to other MBA programs—that is, they cite tier membership rather than their exact rank.[2]

Lessons about perception suggest that you should never, ever be 11th! (Actually I was once 11th, but that is a story for a different book.)

But now we turn to another fascinating situation. "Top 100" also has a magic power among other numbers in the world of higher education. The competition for being in the list of the world's "top 100" universities has transformed the concept of being a "world-class" university into an explicit target of administrators.

Nobody likes it, but everybody uses it: university ranking

A recurring theme in our complex world pertains to the question of whether it is possible to summarize the performance of an organization faithfully with a single score. Universities, colleges, and schools are complex social organizations that serve a variety of purposes, and measuring their performance is obviously delicate. What does it really mean if we say that one university ranks 27th and another one ranks 42nd? How do these numbers influence the decisions made by the big stakeholders of the college ranking game—students and their parents, admissions offices, and college administrators? While university ranking has become an obsession in this century, it is not a strong exaggeration to state that *everybody* simultaneously criticizes and uses rankings. Ranking is and remains with us, so the best thing we can do is to understand the rules of the game. We should keep in mind the lesson hopefully learned by now: ranking reflects a mixture of the reality and illusion of objectivity, and it is also subject to manipulation.

While our obsession with ranking is relatively new, there are early precursors to quantitative analysis of universities. In an isolated, pioneering work published in 1863, a Czech professor at the Prague Polytechnical Institute, Carl Kořistka, analyzed and compared technical universities in leading European countries.[3] The university known today as Karlsruhe Institute of Technology, one of Germany's leading engineering schools, had the largest number of students (about 800) and 50 professors, according to Kořistka's analysis. If we believe the numbers, the student/faculty ratio has been reduced from 16:1 to 5:1, since nowadays the 25,000 students of the university are served by 6,000 academic staff. It is interesting to see that while the foreign student population at Karlsruhe constituted about 60 percent of the student body, at Berlin's institution only 2 percent (7 out of 374) of its students came from foreign countries. The range of student/faculty ratios at the institutions for which Kořistka found reliable data was between 8:1 and 18:1. (Kořistka himself did not specifically calculate the student/faculty ratio, probably because it was not the focus of institutions of higher education at the time.)

James McKeen Cattell (1860–1944) was a pioneering professor in the United States who contributed much to the transformation of psychology from pseudoscience to legitimate science by adopting both experimental and quantitative methods. He was motivated by Francis Galton, among others, who, as we remember, liked to count and measure everything. Cattell was inspired to study distinguished men of science. He asked a number of competent men in each field to rate their colleagues or, more precisely, denote their excellence with stars. He then characterized institutions by the ratio of starred scientists to the total number of faculty and ranked the results. Cattell's aim was to provide help to both potential students and institutions. The first edition of *American Men of Science* was published in 1906 and the seventh by 1944.[4] Cattel's approach suggested that the quality of the universities could be measured by the number of excellent faculty, and it helped lay the foundation for our way of thinking about university ranking. The importance of "distinguished persons" in the ranking procedure ensured the priority of older, private institutions over newer, public universities. Other early ranking systems added several more criteria, including graduates' success in later life, which is an output measure of teaching quality, and volumes in the library, which, along with student/faculty ratios, is an input measure of the resources.[5]

Symbolically, our modern obsession with university ranking is represented with the appearance of the ranking by *U.S. News & World Report* (*USNWR*) in 1983, marking the entrance of mass media onto the scene. *USNWR* simultaneously wished to provide accessible information for students and parents and to increase the visibility and revenue of the magazine. Soon the *USNWR* list became a measure of reputation, and college administrators (not necessarily by their own admission) made it their explicit target to rise in that ranking. The reputation race shifted gears.

USNWR discriminates between rankings for *best quality* versus *best value*. To calculate best value, quality is given a weight of 60 percent of the overall score, the percentage of students receiving need-based grants 25 percent, and the average discount awarded to students 15 percent. *USNWR* changed its methodology in response to criticism and now combines more objective input data (resources, entering student quality) with the subjective aspects of reputation. However, it is difficult to enter a race when the rules are changing.

While the US (and UK) ranking systems were followed by the emergence of many national ranking systems, the race became much more exciting with the appearance of global ranking. The three most influential global rankings are those produced by Shanghai Ranking Consultancy (the Academic Ranking of World Universities), *Times* Higher Education, and Quacquarelli Symonds. However, rather than measuring teaching performance, they place greater emphasis on the research produced by institutions of higher education.

Demand for ranking

Transparency, accountability, and comparability

There is an increased demand for transparency, accountability, and comparability in institutions of higher education from both the public and politicians.[6] Ranking methodology offered a simple and easily interpretable comparison. In her excellent book *Rankings and the Reshaping of Higher Education*,[7] Ellen Hazelkorn published a list of typology instruments of transparency, accountability, and comparability:

- Accreditation: certification, directly by government or via an agency, of a particular higher education institution (HEI) with authority/recognition as an HEI and the power to award qualifications
- Assessment, quality assurance, and evaluation: assesses institutional quality processes, or quality of research and/or teaching and learning
- Benchmarking: systematic comparison of practice and performance with peer institutions
- Classification and profiling: typology or framework of HEIs to denote diversity, usually according to mission and type
- College guides and social networking: provides information about HEIs for students, employers, peers, and the general public
- Rankings, ratings, and banding: enables national and global comparison of HEI performance according to particular indicators and characteristics that set a "norm" of achievement

The different instruments serve the dual purpose of reflecting past performance and helping plan future activity.

Heterogeneity and comprehensiveness

In 2011 Malcolm Gladwell explained the nuts and bolts of college rankings in a *New Yorker* article titled "*The order of things*: what college rankings really tell us." He describes the evolution of the *USNWR* ranking systems and the difficulties of being both "comprehensive *and* heterogeneous" (Gladwell's italics). Comprehensive means that nearly all aspects of something are included, while heterogeneity attempts to account for the diversity in HEIs. Gladwell gives the example that heterogeneity

> aims to compare Penn State—a very large, public, land-grant university with a low tuition and an economically diverse student body, set in a rural valley in central Pennsylvania and famous for its football team—with Yeshiva University, a small, expensive, private Jewish university whose undergraduate program is set on two campuses in Manhattan (one in midtown, for the women, and one far uptown, for the men) and is definitely not famous for its football team.

I think from the example it is clear that comparing these two institutions is much more difficult than comparing apples and oranges. We saw in Chapter 2 that even the latter is quite difficult. As concerns comprehensiveness and heterogeneity, there is a tradeoff between the two characteristics. Thus, ranking universities is a matter of measures. How does it work?

What does ranking measure? Indicators and weights

At a certain point the *USNWR* rankings used seven indicators and weights to assign a single number to each HEI:

- Undergraduate academic reputation: 22.5 percent
- Graduation and freshman retention rates: 20 percent
- Faculty resources: 20 percent
- Student selectivity: 15 percent
- Financial resources: 10 percent
- Graduation rate performance: 7.5 percent
- Alumni giving: 5 percent

The *Times* Higher Education World University Rankings system advertises itself on its website as "the only global performance tables that judge research-intensive universities across all their core missions: teaching, research, knowledge transfer and international outlook. We use thirteen carefully calibrated performance indicators to provide the most comprehensive and balanced comparisons, trusted by students, academics, university leaders, industry and governments." Of course, there are somewhat arbitrarily set weighting factors. As concerns the broader categories, their numerical values are set as:

- Teaching (the learning environment): 30 percent
- Research (volume, income, and reputation): 30 percent
- Citations (research influence): 30 percent
- International outlook (staff, students, research): 7.5 percent
- Industry income (knowledge transfer): 2.5 percent

Create your own ranking

Emanuelle Tognoli, a clever and charming French professor of complex systems and brain sciences at Florida Atlantic University, made a remarkable comment on my blog:

> As we develop computational literacy in the decades to come, perhaps we will adopt "**personalized rankings**" [my boldface],

> just like we do for "personalized medicine": each and everyone will be able to weight the factors (rank = 30% teaching + . . .) and write their own equations (or have a website write it for them with sliders) to see their unique customized rankings depending on their own priorities. This is in effect what cognition tries to accomplish when selecting a University or buying a new computer the plain old way, with the limits we know inherent in the manipulation of high dimensional state spaces. Then the information source or authority would change its role, it would have to spend more time explaining why the factors matter so that the user can make an informed decision when adjusting the weights. Do you think dear Peter that those multitudinous rankings would be as successful and popular as their rigid counterparts? Would they be more/less useful? How will they affect the users (me trying to find a good University)? Human stakeholders in the ranked entities (those Universities)? And the people who commit resource to set up those rankings (Times Higher Education ranking said Universities)?

Europe was shocked by the first results of global ranking and initiated a new project. The slogan of U-Multirank,[8] "Create your own ranking," is close to what Emanuelle suggested. U-Multirank is designed to let students choose what's important to them and find not the best overall school but the best personalized school. Students more often than not do not know their preferences explicitly, so they should answer simpler questions: What should I study? Where do I want to study? The goal of U-Multirank is to create a more flexible system that allows users to choose the most appropriate dimensions of comparison. U-Multirank emphasizes its multidimensionality, integrating research, teaching and learning, international orientation, knowledge transfer, and regional engagement.

Views on U-Multirank, not surprisingly, are mixed. Some feel that it struggles with the comparability and reliability of data. Major concerns have also been raised regarding whether the indicators can be interpreted consistently across different institutions and countries.

We should leave with the fact that no ranking system can capture all aspects of a college or university. Again, we are navigating between subjective and objective. There is no perfect, objective ranking system. A ranking is a subjective opinion about which indicators are significant and how they are weighted to analyze available data.

The halo effect again: the biasing role of reputation

Every year, *USNWR* sends a survey to the country's university and college presidents, provosts, and admissions deans (along with a sample of high-school guidance counselors) asking them to grade all the schools in their category on a scale of one to five. Those at national universities, for example, are asked to rank all 261 other national universities.

Gladwell unfolds a story, in his usual elegant style, about legal experts who ranked a nonexistent law school:

> Some years ago, similarly, a former chief justice of the Michigan supreme court, Thomas Brennan, sent a questionnaire to a hundred or so of his fellow-lawyers, asking them to rank a list of ten law schools in order of quality. "They included a good sample of the big names. Harvard. Yale. University of Michigan. And some lesser-known schools. John Marshall. Thomas Cooley," Brennan wrote. "As I recall, they ranked Penn State's law school right about in the middle of the pack. Maybe fifth among the ten schools listed. Of course, Penn State doesn't have a law school." Those lawyers put Penn State in the middle of the pack, even though every fact they thought they knew about Penn State's law school

> was an illusion, because in their minds Penn State is a middle-of-the-pack brand. Sound judgments of educational quality have to be based on specific, hard-to-observe features. But reputational ratings are simply inferences from broad, readily observable features of an institution's identity, such as its history, its prominence in the media, or the elegance of its architecture. They are prejudices.

The example (and there are many similar ones from Princeton to Heidelberg) is a manifestation of the halo effect. As the reader may remember, our general impressions influence our qualifications of a specific trait.

I think the story might have a positive interpretation. As we see the actual ranking numbers now, Penn State Law Schools (actually, there are now two separately accredited law schools of the Pennsylvania State University) have intermediate ranks, so the experts' prejudice correlates very well with the actual value, and we may see their judgment as the predictive power of the collective wisdom. Comparison does not have any alternative in the global world of higher education. Self-declaration and self-evaluation do not convince students, peers, and other stakeholders anymore. We don't have a single "ideal" ranking system. Even at a superficial level there is a dichotomy between excellence in research and excellence in teaching. However, potential students might be more interested in the quality of their local environment, and an ideal ranking system should evaluate both teaching and research output on a department-by-department level.

Another highly publicized example of the halo effect can be found in a ranking made by the German business magazine *Handelsblatt* in which employers rated business studies programs. They rated Heidelberg University, which generally has a high reputation, among the top six, even though that university did not have a business studies program.

Ranking game: from reflection to reaction

In a highly cited paper and in their book *Engines of Anxiety*,[9,10] Wendy Espeland and Michael Sauder clearly demonstrate that school rankings provide not only a passive mirror but an impetus for change. Rankings have acted, on the one hand, as a warning signal for HEIs, challenging self-perceptions of greatness, by nations, by institutions, and by individual academics. In a global marketplace, it is much better to eliminate self-qualification and self-declaration and replace it with formal international comparisons. On the other hand, ranking is the driving force for reactive changes. Two mechanisms of reflections for ranking have been identified: self-fulfilling prophecy and commensuration.

Self-fulfilling prophecy is a mechanism of self-amplification of even small differences. Even small differences in rank in one year affect the number and quality of applications for the next year. Consequently, selectivity scores will be different, and it has a causal effect in the computation of ranking. Thus, finally it might happen that statistically insignificant measurement noise will make a meaningful difference in the ranking. Old rankings can influence new ones by the biasing role of reputation. It is difficult to imagine that even the most experienced administrators have more than superficial knowledge about the majority of other schools.

Commensuration is a second important mechanism of generating reactive responses to rankings. First, qualities are transformed to comparable quantities. Cost/benefit ratios, prices, standardized tests, etc., we already know these things from everyday life. Commensuration is a framing process, shaping what we pay attention to. Since limited attention is a key facet of human cognitive capacity, commensuration is extremely important. Participants decide what will be the subject of the discussions and what will be neglected. In their excellent book about the politics of attention, Bryan Jones and Frank Baumgartner[11] define what we might see as a two-step process. First is what is called *agenda setting*, which means, at the pragmatic level, the choice of indicators. Those qualities that will not be mapped into numbers will be neglected. Second, our reaction is disproportional: any policymaking systems continuously under- and overreact. At this moment ranking algorithms don't take into account such concepts as "free speech on the campus," the number of gender-neutral facilities, etc. Should we? Commensuration leads to reduction and simplification.

How do the different stakeholders use ranking?

Let's look at two examples. Administrators at about 170 European universities were asked, "Do rankings play a part in your institutional strategy?" The responses were as follows:

- No: 39 percent
- Yes, and our institution formulated a clear target in terms of its position in national rankings: 14 percent
- Yes, and our institution formulated a clear target in terms of its position in international rankings: 18 percent
- Yes, and our institution formulated a clear target for both national and international rankings: 29 percent

I asked a set of my senior undergraduate students how they are using ranking systems to choose graduate schools. Here is one answer:

> I am an undergraduate Computer Science student and am currently applying to PhD programs in Machine Learning. Personally, I heavily referenced the rankings when choosing the schools to apply to and to not apply to. I consulted several different lists, including the "Best Graduate Programs: Artificial Intelligence" list on the U.S. News site. Since there are hundreds of colleges, I wanted a way to narrow my initial search.

So I used such rankings as a starting point. Instead of considering all schools, I considered the top forty or so schools. I then considered location, which is more subjective since I have personal preferences that may not be shared by everyone (i.e., I enjoy the rainy, misty weather of the Pacific Northwest). Then, from this narrowed search, I selected schools who had professors and research labs who interested me. In my experience, the use of rankings were used as a filter to complement my personal preferences in order to select schools that would fit a wide range of criteria I was looking for.

Should we or shouldn't we?

Since 18,000 HEIs can be found in the World Higher Education Database, only 0.5 percent of them can make the "top 100" cut. I am inclined to believe that it is not true that there is only one game in town. While competition is a positive driving force, it is not true that all the universities should go to the same starting line. Newer and smaller universities, especially from emerging economies, generally don't have massive financial and other resources. It is almost impossible for these institutions to improve their ranking status. Of course, there are categories—for instance, *USWNR* ranks national universities, liberal arts colleges, and regional universities and colleges. Society needs the middle- and lower-ranked universities and colleges, but it might be to their benefit to concentrate on helping students earn credentials and employment rather than spending too much to earn a better image in the ranking game. I might be wrong.

Is there any "best" country in the world?

As we all know, humankind has organized itself into geopolitical units called countries. Historically, people have preferred to belong to a particular country and to share a sense of national identity. *Homophily* is an ancient trait: we like to spend more time with people who are like us. While some people believe that the idea of countries, as nation-states, is outdated and the source of conflict, countries remain a primary means of controlling people, organizing society, and managing the distribution of wealth.

Countries are ranked and rated now by an enormous number of criteria, adopted by hundreds of different organizations, sometimes strongly connected to specific countries (frequently to the United States). In a book about the ranking of countries,[12] authors Alexander Cooley and Jack Snyder identify 95 indices that have been introduced to evaluate and compare states. The indices are lumped into categories like "Business and Economics," "Country Risk," "Democracy and Governance," "Environment," "Media and Press," "Security Issues and Conflict," "Social Welfare," and "Transparency."

A ranked list of countries based on the social welfare function defined by Amartya Sen has been prepared annually by using data from the Central Intelligence Agency, and another version is prepared using data from the International Monetary Fund and the United Nations. (Recall that the Sen social welfare function is calculated as product of gross domestic product [GDP] per capita and the difference between one and the society's inequality measure, and it is reported in terms of dollars per person per year.) The last published list is from 2015:

1. Qatar, 82884
2. Luxembourg, 49242
3. Norway, 47861
4. Singapore, 43518
5. Switzerland, 42335
6. Netherlands, 34853
7. Sweden, 34443
8. Denmark, 33907

9. Germany, 33719
10. Iceland, 33695
11. United States, 33260

Qatar has a well-developed oil exploration industry, and the petroleum industry accounts for 60 percent of the country's GDP. Its low (but rapidly increasing due to an influx of migrant workers) population contributes to a large GDP per capita. The population explosion due to the immigration of (young) males has produced an extreme gender imbalance: there are only about 700,000 women in a country of 2.5 million people. Many immigrants, mostly involved in building the infrastructure needed for the upcoming World Cup, live in labor camps. However, since the Gini inequality index measures income inequalities but not social inequalities, Qatar still leads the list.

The scores of the last six countries are close to each other, and the specific ranking does not have too much significance. Even so, it is somewhat remarkable that the United States did not manage to make the top-10 list.

Is a horse bigger or smaller than a cow?

Ferenc Jánossy (1914–1997), an engineer turned economist from a legendary Hungarian family (he was the stepson of George Lukács [1885–1971], one of the founders of the philosophy of "Western Marxism"), wrote a book in Hungarian titled *The Measurability and a New Measuring Method of Economic Development Level*, and it was a revelation at that time. Jánossy explained his approach clearly with a reference to an anecdote about comparing animals:

> The first issue is how qualitatively different objects can be compared quantitatively. Every child knows that an elephant is bigger than a sparrow. They would agree without the least doubt

> that the cow is smaller than the elephant, but bigger than the sparrow. Ranking animals according to size, they would place the cat between the cow and the sparrow without any hesitation. But suddenly the child is faced by the problem of the horse. Where should the horse go? Is it bigger or smaller than the cow? When comparing objects of different characteristics, ranking is no longer so simple because taking into consideration various features may lead to various ranking results. (The horse is taller yet narrower than the cow.)

Generalizing the above game, Jánossy finds that the greater the qualitative difference between two items, the greater the quantitative difference needed to make the ranking reliable according to size. Qualitative difference limits quantitative comparability—this is what Jánossy calls the "criterion of comparability." Obviously, the critical limit depends on the features compared. (If ranking is only according to height, then the horse–cow dilemma does not even arise.) Ranking, however, is not the end but only the means; therefore, the basis of the comparison cannot be changed to make ranking easier. A clear definition of the organizing principle may lower the critical limit but cannot eliminate it.

The next question is how to move from ranking to measuring. How do we make a quantitative statement? Or rather, under what conditions could we quantitatively describe how, for example, Sweden is more advanced than Turkey? If any one feature is not additive or cannot be traced back to some additive feature, it cannot be measured. If a feature is measurable, then the comparison of two objects can be decomposed into two steps of "numerical measurements along a fixed scale," which means that the critical limit of measurability matches the given absolute scale and the limit of comparability of the object. If the examined feature of the objects can be measured along an absolute scale, then the critical limit can be expressed numerically.

This example hints that ranking and rating need appropriate methods, and the methods have limits. It is vital to accept the existence of the limits of comparability.

Pay more taxes and be happier

While there are almost infinitely many ways to rank countries, many readers will agree with me that one of the most important questions to answer is how happy a country is. In 2011, the United Nations General Assembly initiated a project that sought to measure the happiness of citizens of member countries. But how do we measure the happiness of a country? The measurement is mostly based on a simple task: in each country, a significant number of people are asked: "Please imagine a ladder with steps numbered from zero at the bottom to ten at the top. The top of the ladder represents the best possible life for you and the bottom of the ladder represents the worst possible life for you. On which step of the ladder would you say you personally feel you stand at this time?"

A report issued by the United Nations in 2017 ranked Norway as the happiest country in the world. (The reader already knows that the phenomenon of "pecking order" among chickens was discovered in Norway, and the Norwegian Magnus Carlsen has the highest Elo number. The neurobiologists among our readers will also remember that May-Britt Moser and Edvard Moser from Norway were awarded the Nobel Prize in Physiology or Medicine in 2014 for discovering certain types of neurons called grid cells, which are responsible for spatial information processing.) It was remarkable to see the reaction of Prime Minister Erna Solberg: "Even if we top this statistic now we must continue to prioritize mental health care."

Actually there is no statistically significant difference among the happiest five countries, which each received scores around 7.5 (Norway 7.54, Denmark 7.52, Iceland 7.50, Switzerland 7.49, and Finland 7.47). The Central African Republic had the lowest score, at −2.69.

In 2018 Finland took the lead, and the United States ranked 18th out of 156 countries surveyed, down four spots from 2017's report. Despite a strong economy, the United States ranks quite poorly on social measures such as life expectancy and suicide rates. Major factors possibly contributing to this drop in ranking are the worsening of the opioid crisis, the growing economic inequality, and the decrease in confidence in government.

Investment in mental health care is likely to correlate to average happiness. A good proxy for investment in mental health care is the number of psychiatrists and psychologists working in mental health care per capita. Based on these figures, average happiness appears to be higher in countries that invest more in mental health care.[13] Like it or not, developed mental health care implies the more frequent use of antidepressants, and increasing use of antidepressants and decreasing national suicide rates have been reported recently from the happiest country in the world.

The other side of the happiness story is related to suicide rates. Hungary has a history of high suicide rates, demonstrated by statistics stretching back for more than a century.[14] In a majority of years between 1960 and 2000, the suicide rate of Hungary was the highest in the world. I belong to that camp that believes that Hungarians have some problem with their (our) self-identification. We have an arguably isolated language, and the country has never been an emancipated member of the West but has also lost its Eastern origin. There has been some improvement in the last 20 years, though. Interestingly, changes are not directly related to socioeconomic development, as Lithuania and South Korea are now among the "leading" countries in terms of suicide rates.

Happiness and money

Individual ranking reflects just one particular unidimensional projection of our complex world, and it is far more complicated when other dimensions are taken into account. The popular question of whether or not we can buy happiness can be converted into the study of whether there is any correlation between happiness and wealth.

There is an ongoing debate about the so-called Easterlin paradox (named for Richard Easterlin, a professor of economics at the University of Southern California):

- Within a society, rich people tend to be much happier than poor people.
- However, rich societies tend not to be happier than poor societies (or not by much).
- As countries get richer, they do not get happier.

The economists Betsey Stevenson and Justin Wolfers found that Easterlin was simply wrong,[15] and they argue that there is a monotonic increasing relationship between income and happiness. The increase, however, is not linear but logarithmic (please remember, the only math needed to read this book is to know the form of a logarithmic function). The happiness value of the next dollar you earn is always worth less, and it leads to saturation. After many years (mathematicians like to call this point an *asymptotic limit*), the paradox is right. While the Easterlin paradox has been challenged and may not always be supported by data, it is not necessarily wrong to say that we should devote less time to making money and more time to family life and physical and mental health!

Ranking countries by credit rating: the objectivity–subjectivity dilemma again

We already know that individuals get *credit scores*, while corporations and governments receive *credit ratings*. This is just the jargon. Governments of countries require ratings to borrow money. Credit ratings also reflect the quality of a country as an investment target, and a country's credit rating depends on its economic and political state. Why do countries need credit ratings?[16]

> Many countries rely on foreign investors to purchase their debt, and these investors rely heavily on the credit ratings given by the credit rating agencies. The benefits for a country of a good credit rating include being able to access funds from outside their country, and the possession of a good rating can attract other forms of financing to a country, such as foreign direct investment. For instance, a company looking to open a factory in a particular country may first look at the country's credit rating to assess its stability before deciding to invest.

It is well known that the United States leads the list of countries ranked according to external debt, followed by the United Kingdom. It is remarkable that Luxembourg has much larger debt per capita than any other country. Luxembourg is known as a major financial center, so presumably the country owns large deposits belonging to foreign people.

In principle, the rating process should give an objective and independent assessment. If the procedure were totally objective, it would be sufficient to have only one credit rating agency (CRA). But in the United States, we have three big agencies (Fitch, Moody's, and Standard & Poor) and many smaller ones, who might use different databases and (generally private) algorithms, and they therefore produce (slightly) different results.

Capsule history of the three famous CRAs

In 1860, Henry Poor (1812–1905) published *History of Railroads and Canals in the United States*, an attempt to collect and provide comprehensive information about the financial state of such transportation companies. Standard Statistics started to publish ratings of different bonds in 1906, and they merged in 1941 to form Standard and Poor's Corporation. Their product, the S&P 500, became a stock market index, a measure of economic activity. John Knowles Fitch (1880–1943) founded the Fitch Publishing Company in 1913 to provide financial statistics for helping investors to make decisions. In 1924, they introduced the AAA through D rating system that has become the industry standard for bond ratings.[17] John Moody (1868–1958) and his company first published *Moody's Manual* in 1900. Moody's Investors Service has provided ratings for nearly all of the government bond markets and today is a full-scale rating agency.

The Latin phrase "Quis custodiet ipsos custodes?" is literally translated as "Who will guard the guards themselves?" A natural question arises: Who rates the CRAs?[18] In 1975, the designation of "nationally recognized statistical ratings organizations" was created. Investors simply needed more reliable information to help them decide how to allocate their resources, and this demand has led to enormous growth, expansion, and influence of the credit ratings industry. The Credit Rating Agency Reform Act of 2006 allows the main regulatory agency, the Securities and Exchange Commission, to regulate internal credit rating processes. CRAs played a critical role in the financial crisis of 2008, and the details are far beyond the scope of this book. The lesson I learned from Michael Lewis's bestseller[19] was "The line between gambling and investing is artificial and thin."

Criticisms of CRAs based on their subjectivity

The fact that CRAs also perform consulting services is an obvious source of potential bias in ratings. (Remember the story of the wolf, the self-appointed judge of the other animals, whose judgment translated into their death sentence!) The credit ratings game is played under the condition that their principal source of revenue comes from the firms whose products they are rating.[20] CRAs have been accused of biased evaluation and violating principles of objectivity. Generally, CRAs have denied the existence of any conflict of interest. They have stated that rating decisions are not made by individuals but by committees, and the analysts have not received any compensation based on their ratings.

Rating agencies now use mathematical models, the details of which are not fully disclosed. We already know that models are based on human assumptions. Furthermore, the results of any model can be overridden by humans (they might be called a "rating committee," whose activities are kept secret). To make the rating procedure more transparent, the Dodd-Frank Wall Street Reform and Consumer Protection Act (2010) required CRAs to disclose their methodologies. Although some provisions of the law have since been repealed or altered, many of the portions concerning CRA regulation remain intact.[21,22] So we have the question: Who should have the last word, the computer or the human?

Objective algorithms versus subjective conflict of interests

We have again the dilemma: What would be the difference between a subjective and an objective credit rating? A subjective credit rating would be one individual's, or CRA's, point of view. It would reflect the expertise of a particular analyst and her agency's proprietary algorithms. An objective credit rating, on the other hand, would be something based on open databases and open-source algorithms.

If CRAs were really objective, then there wouldn't be any need for more than one agency, and we wouldn't have different rating results. If there were only one such rating, CRAs would not generate such large revenues because anybody could simply use the publicly available, consistent criteria provided by a single rating agency to generate the (objective) ratings. There would be no incentive for CRAs to manipulate rankings in order to generate revenue.

Unhappy reactions: from China to Europe

Several years ago, S&P downgraded China's credit rating, and the finance ministry of the huge country vehemently criticized the validity of both S&P's procedure and its result. Other developing countries, most importantly India, continuously rebel against the CRAs. India has a battle with Fitch, which has refused to upgrade India's credit ratings each year since 2006. Since India has recently tried to attract more foreign investment, it is very painful to get a mediocre grade for creditworthiness.

Europeans generally feel that the "Big Three" CRAs show bias toward the United States. The United States has managed to maintain its AAA rating despite a growing deficit and high levels of public debt. But in August 2011, S&P downgraded the US's credit rating to AA+ for the first time ever in history. The other two agencies still assign top credit scores to the United States, but S&P affirmed the US's AA+ credit in 2018, reflecting the balance between positive and negative factors expected over the next two years.

Should we or shouldn't we?

Despite the debates over the merit of credit ratings, they remain a crucial facet of the international financial system. The spirit of this book is in accordance with the evaluation of Sebastian Mallaby from the Council on Foreign Relations.[23] The best way to counter the monopolistic power of the Big Three, he argued, is for investors to stop giving their ratings so much weight: "The reason why the subprime bubble could happen, or the reason why the European sovereign debt crisis can happen is, largely, that very blind investors bought bonds relying on ratings, and [didn't do] their own homework about what the real credit risk was in the bonds."

Ranking corruption: from toleration to condemnation

Corruption: what, why, and how

The *New York Times* archive gives 162,751 search results for corruption (as of September 19, 2018). Here is a short list of titles, which clearly demonstrates the omnipresence of using public authorities and resources for gaining political and/or personal advantages:

- Corruption in Mexico
- The Full-Spectrum Corruption of Donald Trump
- Former Argentine President's Homes Searched in Corruption Inquiry
- South Africa Vows to End Corruption. Are Its New Leaders Part of the Problem?
- Guatemalan Leader Bars Re-entry of Corruption Prosecutor
- In a Corruption Battle in Honduras, the Elites Hit Back
- How Ukraine Is Fighting Corruption One Heart Stent at a Time
- Can Peru's Democracy Survive Corruption?
- Violence Erupts as Tens of Thousands Protest Corruption in Romania

The most serious form is grand corruption, which according to Transparency International (TI) is "the abuse of high-level power that benefits the few at the expense of the many, and causes serious and widespread harm to individuals and society. It often goes unpunished."[24] In these cases, the international community has the responsibility and obligation to act collectively.

Socioeconomic underdevelopment, ethnic fragmentation, and lack of accountability seem to be significant factors behind corruption. Corruption certainly has a destabilizing effect on the whole political system and, by repelling foreign investors, slows down any economic development. To analyze quantitatively the negative consequences of corruption, measures have been defined.

Measuring corruption

TI, a nongovernmental organization headquartered in Berlin, was established to systematically monitor corruption throughout the world. The group prepares a report annually. Using its Corruption Perception Index (CPI), it assigns each country a score (on a scale from 100 [very clean] to zero [highly corrupt]). CPI is roughly the inverse of what might be called the Rule of Law Index. How are these scores created and how objectively do they reflect reality? TI aggregates the results of surveys and assessments from about a dozen institutions, and the qualitative evaluations are mapped into a single number. I don't think there is a mapping algorithm; if there is one, I am almost sure it is not publicly available.

A somewhat more objective measure of corruption proposed by Miriam Goldman and Lucio Picci[25] is based on the difference between the amount of infrastructure produced and public spending on it. Where the difference is larger between the monies spent and the existing physical infrastructure, more money is being leaked out due to bribes and fraud, meaning that corruption is greater.

A third measure, called the Aggregation Worldwide Governance Indicators, developed by the World Bank, adopts six key dimensions of governance (Voice & Accountability, Political Stability and Lack of Violence, Government Effectiveness, Regulatory Quality, Rule of

Law, and Control of Corruption) to measure corruption. Critics, however, say that the index is being oversold: the World Bank Institute advertises it as providing "reliable measurements of governance," but it is far from being objective.

The objectivity–subjectivity dilemma

According to TI, "Behind these numbers is the daily reality for people living in these countries. The index cannot capture the individual frustration of this reality, but it does capture the informed views of analysts, businesspeople and experts in countries around the world." Denmark and New Zealand lead the last list with scores of 90, and Somalia is at the other end with a score of 10. The largest negative annual change (10) was produced by Qatar (which dropped from 71 to 61), while Suriname increased its score by 9 (from 36 to 45).

There is a demand for the development of objective measurements of corruption,[26] and while it is difficult to believe that the CPI is entirely objective, TI has played an important role in focusing attention on the ubiquitous problem of corruption.

Unhappy reactions

But even as TI has successfully shifted focus toward corruption, it has also faced backlash from individual countries and other stakeholders in the international community. In some instances, countries reject the assessments made by TI. For example, soon after TI published its 1996 report, Pakistan's prime minister, Benazir Bhutto, was forced to resign after loud protests against widespread corruption in the country. Additionally, since Nigeria was ranked 148 out of 180 countries, despite the efforts of President Muhammadu Buhari to restrict corruption, the government accused people associated with TI of falsifying reports. However, TI states that they used nine sources to score Nigeria, and none of these sources was an individual. (Yet another example of this book's recurring theme about the data used for rating and ranking!) TI stated that the data used in scoring and ranking countries are not generated by TI but are obtained from independent sources. The group concludes that "it is unlikely that TI staff or associates can influence the position of a country in the Corruption Perceptions Index."[27]

TI also faces criticism for not emphasizing the corruption stemming from transnational corporations,[28] and TI and the Worldwide Governance Indicators might exhibit some bias in detecting and identifying "non-Western forms of corruption," while some Western forms of corruption are labeled "business as usual." Furthermore, the dominance of metrics like CPI contributes to a trap: "in countries where corruption is deeply embedded, development aid is increasingly made conditional on the implementation of reforms which are impossible to achieve without that aid."[29]

Altogether, TI has played an important but controversial role in shifting the attitudes of corruption from toleration to condemnation.

Ranking freedom

Freedom, specifically political freedom, includes freedom of assembly, freedom of association, freedom of choice, and freedom of speech. In the English language there is some difference between "freedom" and "liberty," but in my mother language we have just one beauteous word ("szabadság"). As the British historian Lord Acton (1834–1902), more precisely John Emerich Edward Dalberg-Acton, 1st Baron Acton, famously said: "Liberty is not the power of doing what we like, but the right to do what we ought." He also stated, "The most certain test by which we judge whether a country is really free is the amount of security enjoyed by minorities." Trivially there is no unique definition for freedom/liberty, since there might be an obvious trade-off between individual liberty versus the interest of a community. In any case, ranking countries based on their freedom, as we already know, requires a measurement process.

Measuring freedom

Freedom in the World, published each year since 1972 by the US-based Freedom House (FH), rates and ranks countries on political rights and civil liberties. Based on their numerical scores in these two dimensions, countries are then classified into three groups: free, partly free, or not free. FH describes how objective and subjective elements are combined during the rating process:[30]

> The analysts' proposed scores are discussed and defended at a series of review meetings, organized by region and attended by Freedom House staff and a panel of expert advisers. The final scores represent the consensus of the analysts, advisers, and staff. Although an element of subjectivity is unavoidable in such an enterprise, the ratings process emphasizes methodological consistency, intellectual rigor, and balanced and unbiased judgments.

We already know that the main methodological question facing FH relates to the aggregation of the different indicators to generate a single number. A closely related question is how reliable the final ranked list really is and how the ranking might be different based on the eventual change of the weights of the individual indicators. FH adopts a three-stage process: scores → ratings → status.

I know I'm repeating myself, but every model is based on assumptions. Here the assumption is that the quantification of political rights is derived from the addition of three factors, and the maximal points are assigned arbitrarily:

Scoring = Electoral Process + Political Pluralism + Functioning of Government

The maximal points possible in each section are set at 12, 16, and 12 respectively. The maximal total point a country may receive is 40, so where does a country fall if its total score is 24 or 33? FH decided to assign ratings to scores: 36 to 40 is 1, 30 to 35 is 2, and so on, until we get down to 0 to 5 is 7. But numbers are just numbers; people need words, too, so FH converts the numbers to words (Table 6.1). More precisely, status words are assigned to the scores based on a combination of political rights and civil liberties, but what I want to focus on here is the interdependence of subjective and objective analysis.

For instance, Russia's score for political rights was 5 out of 40, so the rating was 7, and the combined status was "not free." Since "Regional and local elections are typically manipulated to ensure that the regime's favored candidates win," according to FH's report, the score for Electoral Process was assigned a 0. Political pluralism (and participation) scored a 3. One of the reasons is the underrepresentation of women in politics and government; say, only 3 of 32 cabinet

Table 6.1 From scores to ratings

Scores	Ratings
1.0–2.5	Free
3.0–5.0	Partly free
5.5–7.0	Not free

members are women. Functioning of government scored just 2 out of 12 because, among other issues, there is a serious problem with transparency: "Decisions are adopted behind closed doors by a small group of individuals whose identities are often unclear."

More than just a warning signal: a decade of declines

For the 12th consecutive year, in a large number of countries, measures of democracy show a continuous decline. Turkey displays the largest recent downturn. We know that there is a general concern: when the number of autocratic countries is increasing, even larger regions are destabilized, and violent extremists have much ampler room to act.

To satisfy our love of top-10 lists, here is a new one, which categorizes the 10 countries that have experienced the greatest decline in their overall FH scores:[31]

- Turkey
- Central African Republic
- Mali
- Burundi
- Bahrain
- Mauritania
- Ethiopia
- Venezuela
- Yemen
- Hungary

Hungary has registered the largest cumulative decline among the "Nations in Transit," which is FH's term for the former "communist" countries, because its score has fallen for 10 consecutive years.

Should we or shouldn't we?

Country ratings should not be assumed to be accurate, and we should be sensitive to differences among similar objects. I agree with the assertion[32] that international index rankings are popular but dangerous. I think, however, we need more math and not less. Modern statistical procedures might contribute to better characterizing the uncertainty of the scores. The conclusions might be less sharp, stating, for example, "these 17 countries are realistic candidates to make the top-10 list in this and this category."

Lessons learned: the game is not over

There are ongoing debates about the scope and limits of using metrics to measure the institutional performances of everything from schools to law enforcement to health care organizations. While numerical data are susceptible to manipulation or distortion, this should not be taken to mean that the proper course is to abandon the hope of using datasets to improve social programs and institutions.

While the increase in high-stakes testing leads naturally to an increase in cheating by test-takers, this is not a sufficient reason to replace these tests with more subjective methods of evaluating student progress. Further, ranking algorithms use some "built-in" numbers to reflect the weights of the relevant factors. But different weights lead to different results, and I argue that "personalized rankings" based on the weights specified by the users could help all stakeholders.

Countries are ranked and rated now by an enormous number of criteria, adopted by hundreds of different organizations, sometimes strongly connected to specific countries (frequently to the United States). Even the leaders of those countries, who are unhappy with their score and ranking, are forced to react. But the game is not over, and we are struggling for control over our reputations, as we discuss in the next chapter.

7

The struggle for reputation

From "I don't give a damn 'bout my reputation" to reputation management

Reputation is a key factor for ranking artists, singers, scientists, etc. It is not necessarily compulsory to start this chapter with a story about the role of reputation in popular culture. However, I let myself be seduced by recent news, and I find it interesting here to discuss Taylor Swift's ongoing "Reputation" tours.

The importance of reputation in popular culture has been vastly expanded by the Internet and our ability to broadcast information about our lives on social media platforms to both friends and strangers. We carefully curate the best possible images of ourselves, but we may also be subject to scrutiny and cannot control what others say about us, which may become more important than what we say about ourselves.

The recent popularity of Taylor Swift's studio album *Reputation* is a perfect microcosm of both the importance and uncontrollability of our reputations in a digital age. Born out of a spat between the pop star and rapper Kanye West, *Reputation* addresses the constant negativity that Swift faced from individuals online and the ensuing damage to her reputation and attempts to demonstrate that reputations can be falsified and misleading.

With lines like, "My reputation's never been worse, so you must like me for me," Swift speaks to the distinction between public and private personas that everyone navigates on a daily basis. While

Swift's reputation as an artist defines her celebrity and is, as a result, fundamentally distinct from the reputations of most individuals, her desire to call attention to the ways in which reputations can be misleading or manipulated, especially online, provides a telling example of the difficulties of navigating the subjectivity/objectivity dilemma that we have repeatedly encountered in this text. Swift also garnered attention in political headlines during the congressional midterm elections in 2018 with stories like: "Conservatives are turning on Taylor Swift after she endorsed Democrats," and here is a quantified effect: "Trump 'likes Taylor Swift 25% less' after political post."

When I asked family friend and former coworker Judit Szente about the role of the word "reputation" in these songs, she wrote me back that she prefers Joan Jett's "Bad Reputation" from 1981: "I don't give a damn 'bout my reputation."[1]

While I was somewhat surprised that the concept of reputation plays such an important role in rock music, it is less surprising that Gloria Origgi, an Italian philosopher working in Paris, raises and answers the questions: What does it mean to have a good reputation? What do we lose when we lose a reputation?[2] Our character and our actions shape our reputation, and reputation is a form of currency. Our reputation determines whether or not other people invest in us, buy from us, or give us an award.

Who determines our reputation?

How many friends do you have? Despite what Facebook might suggest, we cannot have a thousand friends. We cannot even have a thousand close acquaintances. The British anthropologist Robin Dunbar has estimated the number of persons with whom we can form stable social relationship: 150, which more precisely means between 100 and 200. When I began my blog, aboutranking.com, and searched my email inbox to decide whom I could easily ask to follow my website, I was shocked. The number was 149 (well, only 60 of them kindly pushed the "follow" button). They are the people who know some of my characteristic features and my actions, so my reputation is based on their perception of my activities. But in a broader sense, my reputation is the collective opinion of everybody else, except myself.

As most people know, it takes time to build a reputation. We all know that a single moment is sufficient to destroy a good reputation. As a quote attributed to Warren Buffett says, "It takes twenty years to build a reputation and five minutes to ruin it. If you think about that, you'll do things differently." Unfortunately, even malicious gossip is sufficient to smash a reputation. Having a good reputation among friends might help, as they may defend you even without your knowing.

The traditional mechanism for constructing a reputation is hierarchical. First, your reputation emerges in the layer of people closest to you, so among your friends, and propagates through layers of more distant acquaintances to the friends of your friends of your friends, etc. Modern media outlets have produced other mechanisms and enabled the emergence of overnight popularity. One of my recent favorite examples of overnight popularity is Baddie Winkle, a grandma who conquered the hearts of the Internet when her great-granddaughter posted her photo on Instagram. She now has millions of followers.[3] Of course, being an overnight sensation does not necessarily imply (positive) reputation. Reputation is social information about the value of a person and her activities. To conduct business, for example, people rely on having a good reputation to communicate their trustworthiness to their clients. It is not enough to be honest; you must also be seen as honest in the eyes of the others. (The cynic inside me suggests that although you need to be *seen* as honest, it is not necessary to *be* honest.)

From indirect reciprocity to the evolution of cooperation

In evolutionary theory, reputation is hypothesized to be an important element in solving the social problems related to human cooperation. Is natural selection (i.e., a spontaneous mechanism) sufficient for developing moral rules of cooperation from the interaction of self-interested players? Political scientist Robert Axelrod[4] has investigated this problem for many years. The starting points of his argument are that (1) biological evolution has successfully taken advantage of altruism and (2) genetic algorithms have used evolutionary principles successfully.

Natural selection is conventionally assumed to favor the strong and selfish who maximize their own utility function. But human societies (hopefully) are organized on altruistic, cooperative interactions. One mechanism that leads to the cooperation of originally selfish people is *indirect reciprocity*. As opposed to *direct reciprocity*, the elementary step ("I'll scratch your back if you'll scratch mine") is replaced by the process, "I help you and somebody else helps me." Indirect reciprocity leads to reputation building, and evolutionary game theory suggests that indirect reciprocity might be a mechanism for evolution of social norms by using the increased reputation. It is easier to cooperate with somebody who has a good reputation than it is to cooperate with somebody whose reputation is bad. Reputation helps trust to emerge among people.

As was mentioned in Chapter 3, Martin Nowak and Karl Sigmund[5] have offered a mathematical model to show that cooperation can emerge even if recipients have no chance of returning assistance to their helper. This is because helping improves reputation, which in turn makes one more likely to be helped. Indirect reciprocity is modeled as an asymmetrical interaction between two randomly chosen players. The interaction is asymmetrical since one of them is the "donor," who can decide whether or not to cooperate, and the other is a passive recipient. However, the result of the decision is not localized; rather, it is observed by a subset of the population, who might propagate the information. Consequently, the decision to cooperate might increase one's reputation. Those people who are considered more helpful have a better chance of receiving help. The calculation of indirect reciprocity is certainly not easy. A cooperative donor would like to cooperate with a player who is most likely also a cooperator and would not like to cooperate with a defector. The probability, q, of knowing someone's reputation should be larger than the cost/benefit ratio of the altruistic act:

$$q > C / B \tag{7.1}$$

The reputation game

The remaining part of this chapter is about the reputation game we play. Some of us are ready to take the "I don't give a damn 'bout my reputation" attitude. Introverted hermits don't necessarily want to spend their time networking with the hope of boosting their reputation. In our success-oriented society, one possible strategy is to try to keep the balance between struggle for reputation and external success with our internal peace. Artists, scientists, and small and big companies are competing for reputation. There are three different dimensions of your reputation: who you are, who you say you are, and who people say you are. The first characterizes your personality and identity; the second reflects your communication strategy and expresses how you would like to be seen (as the cat says, "I would like to be seen as a lion"); the third says how other stakeholders participate in the game and describe you and your activity. The rules of the game are often hidden, and we always have some uncertainty whether or not we should play the game. Should we limit our obsession to reputation-based ranking or should we manage our reputation by any means? Let's see some details!

Digital reputation

In the Internet age we have *digital reputations*. Some of our reputation is expressed by numbers, and this whole book is about discussing the reality, illusion, and manipulation of objectivity. One of my peers (one of the 149) has more than forty thousand scientific citations. He does not need any manipulation; he has both nondigital and digital reputation. When I asked him to follow my website he wrote back: "Your new project sounds very interesting. I don't blog, Twitter, Facebook, etc., but if you want to send along something am happy to comment."

Well, this peer is in my age group, but how about the millennials? The sociologist Eszter Hargittai has studied the online skills of millennials.[6] Her results confirmed what many of us college professors see in the classroom: there is an obvious heterogeneity among the students. There seems to be a correlation between the socioeconomic status of the students and their skill in building their own digital reputations, and there are many students whose only level of skill is being able to post on Facebook without thinking about how that post affects their image. While it's important to tell students that digital reputation matters, it is possible to teach students how to build either personal or business reputations online. I hope it is true that honesty is an essential part of building your online reputation.[7] In 2015 Amazon sued 1,114 people who were paid to publish fake five-star reviews for their products, and in the next years the company sued more sellers for buying fake reviews. Individual people, brands, and companies competing for resources (such as jobs, mates, market share) cannot be successful without building their digital reputation. A strong digital reputation helps to distinguish us from others in the crowd when we apply for jobs or when we build a positive profile for our company online. It is now well known that human resource managers search for the digital presence of applicants during the hiring process. Here is an unranked list of things that human resource managers might identify before they reject an applicant:[8]

- Concerns about the candidate's lifestyle
- Inappropriate comments and text written by the candidate
- Unsuitable photos, videos, and information
- Inappropriate comments or text written by friends and relatives
- Comments criticizing previous employers, coworkers, or clients
- Inappropriate comments or text written by colleagues or work acquaintances
- Membership in certain groups and networks
- Discovered that information the candidate shared was false
- Poor communication skills displayed online
- Concern about the candidate's financial background

The measurement of reputation

Lou Harris (1921–2016) developed and applied methods for measuring public opinion. He not only passively measured the "social temperature" of voters and consumers but also offered communication strategies concerning how candidates should change the focus of their attention toward issues of interest to voters. Famously, he worked as a campaign strategist in John F. Kennedy's team in his 1960 presidential race. His company, called now Harris Poll, introduced the concept of *Reputation Quotient* (RQ), which has quantified the corporate reputation, and the score serves as the basis of the annual ranking of the 100 most visible companies. RQ is based on six dimensions of corporate reputation (emotional appeal, products and services, vision and leadership, workplace environment, financial performance, social responsibility), which leads to the selection of 20 variables. The ranking process consists of the nomination phase and the rating phase. In the nomination phase, the best 100 companies are identified by asking several thousand (4,244 in 2017) US adults, while in the rating phase 25,800 participated. As always, there are some arbitrarily chosen elements used to calculate the score. The maximum RQ score is 100. What we see is that from the people's opinion about the qualities of the companies a score is generated. But a number is a number is a number, so we can assign verbal characterization. Actually the RQ performance ranges are as follows:

80 and above: Excellent
75–79: Very Good
70–74: Good
65–69: Fair
55–64: Poor
50–54: Very Poor
Below 50: Critical.

Amazon is in the top spot for the third consecutive year (2016–2018) in the Harris Poll RQ, but there is a somewhat new phenomenon emerging: "Supermarkets are the new superstars in corporate reputation." If you are unsure about judging changes in the sociopolitical atmosphere, you can get a much more comfortable sense by visiting your local grocery store—a store you know and trust, even though it serves Democrats and Republicans. In 2018, four grocery chains (Wegman's, HEB Grocery, Publix Super Markets, and Aldi) made RQ's top-10 list. Two giant tech corporations fell in the same year, since they failed to release a new sensational product, unlike in years past. As the patient reader already knows, these kinds of measurements contain subjective elements, so there are different methods and results. In contrast to RQ's results praising Amazon as number one, *Forbes* reported that the Reputation Institute's analysis declared Rolex the leader of the last three years. The Swiss watchmaker combines the image of constancy and change—constancy, since the appearance of a new product is not remarkably different from that of a similar one produced 50 years ago, and change, since the company has been a forerunner of innovation. The company is known for, among other things, developing a world-class waterproof wristwatch and the first wristwatch with an automatically changing date.

Now we will turn to reputation-driven people: scientists and artists. People in these communities play the ranking game, as we will review now.

Ranking games that scientists play

Rating scientific journals

The publication process

As everybody knows, scientists publish the results of their research in scientific journals. William Shockley (1910–1989), the Nobel Prize winner and co-inventor of the transistor, analyzed and revolutionized how we think about scientific productivity.[9] Shockley explained that, to publish a paper, one must (1) have the ability to select an appropriate problem for investigation, (2) have competence to work on it, (3) be capable of recognizing a worthwhile result, (4) be able to choose an appropriate stopping point in the research and start to prepare the manuscript, (5) have the ability to present the results and conclusions adequately, (6) be able to profit from the criticism of those who share an interest in the work, (7) have the determination to complete and submit a manuscript for publication, and (8) respond positively to referees' criticism.

Philosophical Transactions of the Royal Society is known to be the first journal in English devoted purely to science. Scientific results are not trivial, so a natural question follows: How do journal editors ensure that they are publishing original papers containing real and significant results? Academic journals are published now mostly by commercial academic publishers, and they implement the "peer review" system to control the quality of publications.

The publication process is quite complicated. Academic journals generally have an editor-in-chief, who first receives all submissions to a specific journal. She assigns the manuscripts to one of several dozen associate editors of the journal. The responsible associate editor generally identifies two appropriate reviewers, who suggest what the journal's response should be ("accept," "revise," and "reject" are generally the main categories of response). The peer review system was based on the moral assumption that reviewing a paper is an honorable task, but this is not necessarily the case anymore. People nowadays are busy, as the reader knows, and it has become more and more difficult to find people willing to spend their time reviewing other people's work without getting any compensation. More often than not, papers should be revised based on the suggestion of the reviewers, and, if finally the editors are satisfied, the paper will be accepted. When her paper is accepted, the simple-minded scientist is happy that the long fight with reviewers is over and that her paper will be published and her reputation increased. Scientists rely on their professional reputation in applying for PhDs, grants, promotions, etc.

After the paper is accepted, she receives a letter from the publishers. Nowadays there are generally two options mentioned in the letter:

1. The publisher will print the article (or recently not even print, just upload it to the journal's website). The scientist sells the publisher the copyright, so the publisher owns her work. This means that the publisher keeps the right to sell the paper, and the author won't see any of the profits.
2. The author has the right to pay the publisher to print/upload her work, which the public can then download for free. This option is hypocritically called "open access."

The whole publication system is in crisis, and while free access is the future, it is not clear who should pay the bill, as we have all heard the truism "there's no such thing as a free lunch."

Reputation by citation

As we not-so-elite scientists painfully know, many papers get very few, if any, citations, and a small fraction gets the majority. Analysis of papers published in *Physical Review* (353,268 papers and 3,110,839 citations from July 1893 through June 2003) indicates the following:[10]

- 11 publications with >1,000 citations
- 79 publications with >500 citations
- 237 publications with >300 citations
- 2,340 publications with >100 citations
- 8,073 publications with >50 citations
- 245,459 publications with <10 citations
- 178,019 publications with <5 citations
- 84,144 publications with 1 citation

Thus, we can see that the distribution of citations is very skewed, with just 11 papers receiving over 1,000 citations and the vast majority receiving less than 10. Technically, citations (to but not from a publication) can be described by a power-law age distribution.

Citations, impact factor, and reputation

Scientific papers cite the results of previous research. (As Newton famously said, "If I have seen further it is by standing on the shoulders of Giants.") There are many thousands of scientific journals, but of course they differ from one another both in their topics and in their reputations. Eugene Garfield (1925–2017) was the pioneer in the field of scientific communication and information. In 1955 he published a paper suggesting a citation index for scientific activity.[11] Garfield created a metric for characterizing the efficiency of scientific journals by introducing the concept of *impact factor* (IF), defined as the average number of citations received by recent articles published in a particular journal. IF is calculated as a fraction, where the numerator is the number of citations generated in the current year by items published in the previous two years, and the denominator is the number of substantive articles and reviews published in the same previous two years. The choice of two years was a compromise between giving greater weight to rapidly changing fields and measuring historical influence. The IF is used to compare journals in the "same discipline": for example, mathematicians and cell biologists have very different citation cultures, so any direct comparison would not be useful. Researchers compete to publish in more influential, "higher-impact" journals as a means of increasing their reputations. The IF of the journals in which a candidate has published is a criterion used when making decisions about tenure appointments, promotions, and grant awards. Everybody in the academic world knows stories about tenure denials based solely on the low IF of the academic journals in which an individual published.

Pushback to this approach has begun to surface in recent years. The San Francisco Declaration on Research Assessment (DORA) intends to halt the practice of correlating a journal's IF to the merits of a specific scientist's contributions. According to DORA, this practice creates biases and inaccuracies when appraising scientific research. DORA also states that the IF is not to be used as a substitute for a "measure of the quality of individual research articles, or in hiring, promotion, or funding decisions."

In a fine article in the *New York Times*, Amy Qin explained how the struggle for scientific reputation can drive scientists to publish fake research.[12] Quantitative measures, specifically IFs, play the main role in career promotions (which I believe is still much better than making promotions based on political loyalty). In June 2017, Sichuan Agricultural University in Ya'an awarded a group of researchers about $2 million in funding after members got a paper published in the academic journal *Cell*, which has a stellar IF of 30. I would like to believe that a journal with such a well-deserved, long-term reputation still has a reliable peer review system. (We, editors of journals with much lower IFs, know very well how difficult it is to find reliable reviewers.)

I think the key portion of the article is this: "In America, if you purposely falsify data, then your career in academia is over," Professor Zhang said. "But in China, the cost of cheating is very low. They won't fire you. You might not get promoted immediately, but once people forget, then you might have a chance to move up." I asked the opinion of a peer of mine from the neural network community, De-Shuang Huang, from Shanghai:

> Let me share a few thoughts on rankings. In China, the impact factor of the candidate papers is really important for the academic promotion, grant awards, etc. Because it shows whether the candidate is capable of getting academic promotion or experts in related fields and ranking can provide a standard of reference. While some fraud scandals happened in many countries, such as USA and Japan, somebody may ask: is it probable that it is much more frequent in China or it is the biased perspective of the West? In my view, China's Internet media has developed rapidly. The scale of Internet users is very big. So once the fraud scandals appear, it will immediately spread around. It does not mean that scandal happens a lot in China. I think that's a one-sided view. One might think that the consequences of the uncovered frauds, so the scandals, have much minor consequences in China than in the US. It was clear to me that it's impossible in China. As mentioned above, China has a huge well-developed Internet social media. If the scandal is revealed, the consequences are very serious and irreparable.

The bad news is that fraud techniques across the whole world are becoming increasingly sophisticated. Perhaps banning individuals or organizations that falsify research from participating in the scientific game for a number of years might have some deterrent power, but still I might be too optimistic.

The metrics we (don't) trust

IF seems to be now a fading superstar. As it always happens, alternative metrics have been defined to characterize the prestige of a journal. This prestige depends on the combination of at least two factors, the numbers of citations and the reputations of the citing agents. Of course, the larger the number of citations, the larger the prestige. However, similarly to the PageRank algorithm, a new measure of the scientific influence of scholarly journals, called *SCImago* Journal Rank, takes into account that citations coming from more important journals provide more prestige.

The diligent reader will remember Campbell's law, which is so important that I am repeating it here: "The more any quantitative social indicator is used for social decision making, the more subject it will be to corruption pressures and the more apt it will be to distort and corrupt the social processes it is intended to monitor." Thus, some editors-in-chief may try to manipulate the prestige measures of their journal by requiring specific tasks from their associate editors. An editor-in-chief may write to the associate editors something like this:

The specific requirements from the editorial board for each of its members are as follows:

- Contribute at least one high-quality paper to this journal per year;
- Review at least one submission per year;
- As a reviewer or editor for other journals in the same field, recommend relevant authors to cite our journal papers in their work;
- Cite at least five of our journal papers in your own publications each year;

- Make other necessary exposure, publicity, and recommendations of our journal.

It is matter of taste and a question of personal choice. I don't believe that any IF manipulation is healthy. I like an editorial published in the journal *Research Policy*, written by Ben Martin, a British professor of science and technology and policy studies at Sussex. My own perspective is close to one of his conclusions: "Where the rules are unclear or absent, the only way of determining whether particular editorial behaviour is appropriate or not is to expose it to public scrutiny."[13]

Another emerging alternative metric is called Altmetrics. The basis of this metric is the assumption that online reflections like news articles, blog posts, tweets, and Facebook, LinkedIn, Reddit, and Google+ posts also matter to the reputation of a journal or article. You already know what I will ask: How, *how*, HOW is a score calculated so as to take into account the different factors and Internet resources? Where are the subjective, arbitrary elements? You already know the answer: a significant source of subjectivity is in the choice of weights assigned to each factor. It is easy to believe that a newspaper story is more likely to bring attention to the research activity than a tweet from your friend. The weights of factors considered by Altmetrics are currently as follows:

News: 8
Blogs: 5
Twitter: 1
Facebook: 0.25
Reddit: 0.25
Patents: 3, etc.

I am inclined to believe that as people place less trust in institutions and in experts, populist metrics might become more credible. Altmetrics better measures the public's response to research than the pure academic impact. There are some initial statistical studies regarding whether or not there is a correlation between citations and Altmetrics scores.[14] Probably, Altmetrisc scores should not replace traditional metrics but should serve as an additional measurement of the social impact of research. I have mixed feelings about Altmetrics, but I am reluctantly ready to accept that the crisis of the traditional peer review–based publication system implies that postpublication reactions matter more than they used to. Of course, the majority of academic articles do not get mentioned in the nontraditional sources that Altmetrics covers. (The situation is even worse: maybe half of the papers are never read by anybody; maybe by some editors, if they are not totally overloaded.) There is an obvious discipline-dependent bias: the majority of the top 100 papers are related to biomedical and health issues.[15]

Rating scientists: the objectivity–subjectivity dilemma again

The reader remembers that James Cattell in the early 20th century popularized the idea of systematically ranking scientists. He asked the experts to make the rankings, so the result was amply subjective. Since then, modern indicators have emerged with the hope of identifying more objective measures. There are now online databases such as the Web of Science from Clarivate Analytics (formerly Thomson Reuters), Scopus from Elsevier, and Google Scholar, which help with objective analysis. Mining these databases helps pinpoint the impact of individual scientists. (As always, we should be careful, since the results depend on the database being used.)

So we have the question: Who is the most influential scientist? Maybe somebody who has written many, many papers. Well, this is partially true, but how we do we account for papers that don't generate any attention (perhaps they were published in journals with low IFs)? We might guess that a paper that receives a large number of citations matters more than a paper that receives few or no citations. But we also want to take into account whether or not a given scientist shows consistent activity over the years. Jorge E. Hirsch, an Argentine American professor of physics, introduced a measure of scientific activity that combines the number of articles a scientist has published and the number of times each of those articles has been cited. It is called the *h-index*. For example, if you have 10 papers that each have at least 10 citations, but you don't have 11 papers with at least 11 citations, you have an h-index of 10. Of course, this example is somewhat arbitrary, so speaking a little bit more technically, a scientist has an h-index equal to H if the top H of her N publications have at least H citations each. Of course, as we have demonstrated many times in this book, an indicator like this is a construction, and we should interrogate the scope and limits of these indicators. How useful are they? Hirsch himself, 10 years after the introduction of the h-index, reevaluated it, and I have taken the liberty of copying a longer comment from him here:

> I think it plays a useful role as an "objective" element in the evaluation and comparison of different scientists, complementing other elements that may be more "subjective" such as prestige, peers' opinions, etc., and others that may be less indicative of individual quality, such as the institutions the scientists belong to or the journals in which they publish their work. In the past, it was easier to argue that a scientist was "excellent" without much solid evidence. Now, if a scientist with a low h-index is argued to be "excellent" it is legitimate to ask for an explanation for why the h-index is low: there may or there may not be plausible reasons. Conversely, in the past it was easier to ignore scientists having wide and large impact but not a highly visible "home run." I believe that considering the h-index should result in better decisions pertaining to hiring and promotion of scientists, granting of awards, election

> to membership in honorary societies, and allocation of research resources by agencies that have to decide between different competing proposals. As long as this index is well used I think it should contribute positively to the progress of science and help reward those who contribute to such progress more fairly.

The obsession with metrics has induced the creation of a big industry elaborating many variations on the h-index.[16]

Should we or shouldn't we?

Scientific metrics can be used for good or ill. A major problem with metrics is the well-charted tendency for people to distort their own behavior to optimize whatever is easily measured (such as publications in highly cited journals) at the expense of what is not easily quantified (such as the quality of teaching).

What looks clear now is that, in the overwhelming majority of cases, the result of the game "Who is a more influential scientist" is decided in very early phases. Imagine a soccer game in which the score is 2–1 in the early stages of the game. Then imagine a rule that says the probability of scoring again is proportional to the number of goals already obtained. The final result might be as insane as 45–3. The score of this soccer game is comparable to metrics describing the impact of specific scientists—the better you do early on, the better you are likely to do in the future.

Predicting success

In one of his bestsellers (*Outliers: The Story of Success*),[17] Malcolm Gladwell asserted that success needs both opportunity and the investment of time. He described two famous examples illustrating that 10,000 hours of practice in developing a specific skill is a precondition of succeeding at that skill. The first example was the Beatles, the most popular rock band of all time, who played all-night shows in Hamburg. The second example was Bill Gates, second on the list of the world's richest men in 2018, who had the chance to spend years with computers in his teenage years (in a time when other children lived in the age before computers and phones).

Albert-László Barabási, one of the most successful scientists of our time, is the right person to search for the mechanism of the emergence of success. His new book[18] appeared several weeks earlier in Hungarian than in English, and I was in Budapest on the day it was published. I read the book in the next 24 hours almost without taking a breath. In the introduction he states upfront: "Success isn't about you. It's about us." Specifically, with well-documented scientific citation data available, it was natural for Barabási to study the quantitative laws behind long-term success in science.[19]

If you want to make predictions, you need a model. Models, as we already know, are based on assumptions. The Barabási group assumed that there are three independent mechanisms that contribute to the emergence of success:

1. The amplification of originally slight differences: Papers with more early citations had a better chance of being cited again than papers with fewer early citations. Barabási's initial world fame derived from the huge number of citations generated by his discovery of the mechanism of "preferential attachment" in the evolution of the World Wide Web, as we mentioned in Chapter 4.
2. The aging effect: The novelty of a paper fades, and what were once new ideas are incorporated in later works.
3. Fitness: This tricky mechanism helps to ensure that latecomers also can be successful.

In *The Formula*, Barabási identified five laws of success, the third of which involves predictive power: future success can be calculated if previous successes are multiplied by the fitness. I am applying this law in order to predict the success of *The Formula*. We know much less about the fitness of ranking, so we will see! But we know from Churchill that "success is not final, failure is not fatal: it is the courage to continue that counts."

Rating and ranking artists

It would be difficult to rate artworks based on their aesthetic values. However, around 1930, George D. Birkhoff (1884–1944), a leading American mathematician, introduced an aesthetic measure,[20] defined as the ratio between order and complexity. The complexity is roughly the number of elements that the image consists of, and the order is a measure for the number of regularities found in the image. While many versions of this measure have been introduced since then, mostly based on information theory, mathematicians have been sufficiently clever to accept that a mathematical theory would not be able to grasp the complexities of the aesthetic experience. While Birkhoff knew very well that his measure totally neglects the emotional and intellectual responses that an artwork induces in viewers, still, intuitively everybody feels that Impressionism grasped better complexity than academic painting. It is remarkable story how the transition to Impressionism was governed by an emergence of new business model, which led to a new market-driven rating and ranking of artists.

The changing reputation of painters: from salons to markets

Salons, the Exhibition of Rejected Art, and the emergence of Impressionism

Historically, the *Académie des Beaux-Arts* dominated French art and controlled both its content and style. Religious and historical themes and portraits were supported, but landscapes and still lifes were not really permitted, and precise brushstrokes characterized the style. For centuries, showing at the Salon was a necessary condition for establishing an artist's reputation and career in Paris.[21] A success in the Salon implied the emergence of reputation, both in terms of prestigious jobs (like teaching positions at the *Ecole des Beaux-Arts*) and awards (like the Legion of Honor, created by Napoleon and maintained by all French governments). The selection process was led by the Salon's jury, controlled by members of the Academy. As it often happens, selection committees like the Salon's jury attempt to conserve the status quo. We cannot blame them, for it is due to human nature, but the reputations of artists depended on the institution. As Gustave Courbet (1819–1877) famously stated after the refusal of all the paintings he submitted to the Salon of 1847:

> It is bias on the part of the gentlemen of the jury: they refuse all those who do not belong to their school, except for one or two, against whom they can no longer fight, such as MM. [monsieurs] Delacroix, Decamps, Diaz, but all those who are not as well known by the public are sent away without a word. That does not bother me in the least, from the point of view of their judgment, but to make a name for oneself one must exhibit, and, unfortunately, that is the only exhibition there is.

The Exhibition of Rejected Art (*"Salon des Refusés"*) was established in 1863, as both something of a consolation prize and an alternative pathway to exhibition for artists excluded by the Salon's selection committee. That year, 1863, is referred to as the year of the birth of modern art, as Edouard Manet (1832–1883) exhibited his then-infamous painting *Le Dejeuner sur l'herbe*. (1863 is also the year when the Football Association was founded in England. Football means soccer, of course, in the British context. For Americans, 1863 is the date of the Battle of Gettysburg, turning point of the Civil War. So, modern art, modern sport, and the modern United States were born synchronously.) The breakthrough happened in 1874, when the first Impressionist exhibition was organized. Claude Monet, Edgar Degas, Pierre-Auguste Renoir, Camille Pissarro, and Berthe Morisot called themselves the Anonymous Society of Painters, Sculptors, Engravers. The paintings were modern—still lifes and portraits, as well as landscapes, painted using small, thin, and still-visible brushstrokes. After 1874, Paul Gauguin, Georges Seurat, and other major artists had the opportunity to gain a career without debuting at the Salon. There was a life outside the review of the Salon's juries!

The emergence of market-driven reputation

Paul Durand-Ruel (1831–1922) had a reputation for discovering Impressionists. As one anecdote describes, "One of his artists came in one day with a young French painter, introducing him and saying, 'This artist will surpass us all'—and that artist was Claude Monet."[22] He made an innovative (which also means risky) business by buying a huge number of paintings made by artists with very little reputation. Durand-Ruel was also an early adopter of the single-artist exhibition, called at that time a "one-man show." He also established a journal to explain and support what later became nothing less than modern art. He was not an art historian, but a businessman and an art dealer. His instinct, however, led him to trust and to invest in a totally new school of painters. Soon the Impressionists had won initial reputations at their independent exhibitions, and Durand-Ruel bought between 1882 and 1884 a large number of paintings from Monet, Pissarro, Renoir, and Sisley.

A new mechanism, the *dealer-critic* system, produced a new social market and gradually superseded the academic system. Art galleries emerged to become the forum for modern art to meet its public. But the slowly growing reputation of the Impressionists was not sufficient to ensure artistic *and* financial success. Durand-Ruel made another innovative step: he made the market global by organizing exhibitions in London and New York, in addition to his exhibitions in Paris. In 1886, Durand-Ruel produced an exhibition of 289 Impressionist paintings at American art galleries in New York. The American public was fascinated by the paintings of Monet, Renoir, and others. Many of the artworks sold became the core of Impressionist collections in major American museums. The artistic and financial success obtained with the help of American collectors allowed Durand-Ruel to get out of debt.

The reputations of the Impressionists, the first real modernists, were quickly established in the advanced art world. In the early 20th century the number of "for profit" art galleries grew enough to create a genuinely competitive market. The transition from the monopoly of the Academy to a market-oriented contemporary art market was complete.

Quantifying artistic success

Does the number of artistic reproductions play a similar role to that of scientific citations?

Before the age of Big Data (so 20 years ago, when Google's PageRank algorithm was born), David Galenson came out with a witty idea that the importance of artworks is reflected by the number of illustrations found in 33 art history textbooks. At the top on the list is *Les Demoiselles d'Avignon*, a 1907 painting by Picasso with 28 illustrations, followed by Vladimir Tatlin, a Soviet painter and architect, whose legendary plan for the Monument to the Third International was shown in 25 books. The statistics are very different when given in terms of Google search hits: 158 million and 230,000, respectively.[23]

How should we rank the three greatest modern American painters—Jackson Pollock, Jasper Johns, and Andy Warhol? The number of total textbook illustrations and the number of Google hits show some rank reversal: Jackson Pollock, 135 and

27.4 million; Jasper Johns, 124 and 12.4 million; and Andy Warhol, 114 and 48.6 million.

The highest price ever paid for a Warhol painting was $105 million for a 1963 canvas titled *Silver Car Crash*, while Johns's *Flag* went for $110 million and Pollock's *No.* 5 became the world's most expensive painting when it was sold privately for $140 million. But these are just numbers. As concerns the prices of these paintings, I should cite Barabási's second law: "Performance is bounded, but success is unbounded." Recent quantitative analyses of the prices of artworks support this view. Analysis on data related to the top ten thousand prices of any artwork traded in any contemporary auction worldwide in the last 30 years showed what we now might expect: the 80/20 percent law works. The distributions of art prices in different artistic periods (Italian Renaissance, Dutch and Flemish paintings around 1600, art auctions in London and Paris in the 1800s) show a large deviation from the bell curve and are described by the most famous skew distribution—the power law (actually, a cubic power law in this case).[24] By and large, the prices of the works of a few innovative artists are much higher than those of their followers.

The reputation of artists now depends on the interplay of different types of players: private collectors, corporations, galleries, and auction houses. The economics of the contemporary art is now a huge field of study,[25,26] and quantitative analysis helps to uncover the mechanisms of how artists' reputations emerge.

Toward a network theory of art

Visual art is not the same as it always has been. While the entire history of the Paris Salon over more than two centuries is composed of less than 130,000 artworks, according to one report,[27] more than 350 million pictures are uploaded by Facebook users every single day. Well, an uploaded image is not an artwork, and still there are many artists who compete for well-visited wall space.

Artists are part of a heterogeneous network made up of other artists, gallery owners, dealers, art consultants, art critics, auction organizers, museum professionals, and other players. The reputations of new painters are built up through their associations with one another. If a newcomer is associated with a famous painter by belonging to the same gallery, this causes reputational gains. Again, with the words of Barabási: "Success isn't about you. It's about us." The trajectory of an artist's career occurs on a set of stages, and it is difficult to make the transition from one level to the next. Typically, these stages are (1) university art spaces; (2) small city galleries; (3) major gallery shows in powerhouse art cities; (4) retrospective exhibitions; and (5) "league A" museum shows.

An important initial step is to find galleries. Are there any rules concerning how galleries select the artists they represent? Without a doubt they are looking for artwork with the hope of increasing their success as a gallery. Success can mean different things to different galleries, and it is not necessarily measured by sales numbers. A more local, academic gallery serves to generate community interest and publicity in addition to financial success.

The art market is a collective game, so feedback from visitors, art dealers, gallery owners, and collectors should be very important. Susan Hiller, an American British painter, advises that you don't have to engage directly in the struggle for reputation:

> To a young artist, I would say: just go day by day and see what happens. Don't worry about other people's judgment. If it resonates, then listen, otherwise pay no attention. Self-doubt is always present for artists because we have the job and the privilege of defining problems and then asking ourselves whether we have solved them.

This strategy, however, will not necessarily lead to reputation, and there is a matching process between artists and the main mediators, galleries. Galleries are involved in selling artists on the primary market, and they are professional intermediaries between fresh works of art and their potential viewers/buyers. They are also advocating for artists, who are new players, positioning them on the market and helping them establish reputations. Galleries also have different influences, of course, and some galleries are more powerful than others. One-third of solo shows in US museums go to artists represented by just five galleries. Those same five dealers—Gagosian Gallery, Pace, Marian Goodman Gallery, David Zwirner, and Hauser & Wirth—now dominate contemporary art for rich people and institutions. Hauser & Wirth just opened (in December 2018) a gallery in the famous alpine resort town of St. Moritz, Switzerland. The quantitative law we have mentioned several times already might be valid for auctions, too: Christie's and Sotheby's constitute 80 percent of the entire international fine art auction business.

There are three laws for increasing your reputation, if you feel it is important:

1. Work hard.
2. Spend 80 percent of your time with marketing and 20 percent with your primary activity.
3. Try to find out how to reconcile the first two laws!

The calculation of top artists

Who is a top artist? Like it or not, the top artist is determined and defined by the community. A popular website in this discussion is ArtFacts.Net (AFN), which was established in 2004. We already know that every ranking system needs a database and a ranking algorithm, and AFN uses a database about exhibitions since the *Salon des Refugés* began in 1863 in Paris. The message of the director of AFN, Marek Claassen, is very clear: only connections and visibility matter, so the algorithms used by AFN prioritize these features:[28]

> We introduced a quantitative method to measure how much an artist is embedded in the international art world. We start with long term relationships between artists and galleries or collections that represent them. These are very strong commitments that last very long. We count the number of countries and the number of collections and galleries. And then we look at solo and group shows. The more international artists a gallery or a museum has, the more its exhibitions' value. Let's say that we have an institution like Tate Modern where thousands of artists are collected. If you have a solo show there you get all the points from these artists and your rank will go up extremely. Biennials, group shows work like collections, their value is based on the artists whose works they show. So if there's an Andy Warhol its value goes up a lot.

When I posted this view to my blog, aboutranking.com, John, my mathematician friend (who, as you remember, never had a car) commented: "Outrageous, scandalous! You are good friend—you are good artist? So clearly does it go?" John, while we know that attempts to quantify the aesthetic value of paintings have a long history (using sources as varied as information theory, fractal theory, etc.), like it or not, fame is a commodity we can quantify. Fame is the product of the collective wisdom of the stakeholders of the art market.

Actually, AFN recently announced a change in its main ranking algorithm.[29] The changes were due to dissatisfaction at the static nature of the ranking list—old (and dead) men like Andy Warhol and Pablo Picasso dominate the list. In response, two changes were made. First, a depreciation factor was introduced: older exhibitions should have less impact on today's career. Second, the ranking algorithm was updated so that exhibitions over five years ago will not have any impact on today's ranking. We may justify that ranking reflects business somewhat indirectly, since AFN does not use financial data in its ranking algorithms.

Ranking of artists

So far Warhol and Picasso still lead the list, followed by Gerhard Richter, who has been involved in the movement called "capitalist realism," an open allusion to "socialist realism," the predominant art style in the Soviet Union. The highest-ranking female artist is the photographer Cindy Sherman, who ranks sixth overall. There are 15 females on the top-100 list, and on this list there are 65 living artists.

The youngest person on the top-100 list (actually 62nd) is Kader Attia, who was born in 1970. This contemporary artist spent his childhood in Paris and Algeria. As a result, his formative years involved both Arab and European ways of thinking, and his navigation among the "Christian Occident, the Islamic Maghreb, and the Jewish Algerian Sephardic world" shaped his worldview.[30] He spent several years in the African country Congo and also in Venezuela in South America, so he has obtained a very broad international perspective. Attia received his education in Paris and Barcelona in the 1990s, and now he lives and works in Algiers, Berlin, and Paris. His extremely multicultural background helped him to develop a unique concept, which seems to be the organizational principle of his artistic activity: the notion of repair. The pain of the past, trauma, and its repair, manifested in the scar, are recurrent patterns in Attia's artwork.

No doubt, Attia had the opportunity to become a successful artist. His first solo exhibition of photography took place in 1996 in Congo, which is not necessarily the best stepping stone to success. I am just guessing that his solo exhibition in the galleries of Paris (Martine et Thibault de la Châtre, Kamel Mennour) benefited him and allowed him to enter the world of the Venice Biennale in 2003. Later, his room-sized installation, *The Repair from Occident to Extra-Occidental Cultures*, made him a celebrity.

Attia has received several awards, including the Marcel Duchamp Prize (2016) and the Joan Miro Prize (2017). He has already presented in group shows at MoMA, New York; Tate

Modern, London; Centre Pompidou, Paris; and the Solomon R. Guggenheim Museum, New York. It is the working hypothesis that he will have solo exhibitions in these museums, and he is still going up on the list of the top 100 artists. There is one reason for doubt: Attia does not have an English Wikipedia site, only French and German. Can your story be a global success without the support of English Wikipedia? The only downside for Attia may be that American college students are unlikely to write essays about him.

Nobel and Oscars: the candidates and the winners

Nobel Prize

One important signifier of reputation is the receipt of a prominent award or honor in one's discipline or community. It is difficult to deny that the Nobel Prize can be identified with being ranked as #1. The Nobel Prize takes its name from Alfred Nobel, a chemist, inventor, engineer, entrepreneur, and author who invented dynamite and held 355 patents during his lifetime. In his will, Nobel set aside his fortune to fund prizes for outstanding achievements in physics, chemistry, physiology or medicine, literature, and humanitarian or peace-related work. Since 1901, the Royal Swedish Academy for Sciences, the Karolinska Institute, the Swedish Academy, and a committee elected by the Norwegian Parliament have been responsible for awarding the Nobel Prizes, and in 1968, the Sveriges Riksbank established a prize in economic sciences that is awarded alongside the Nobel Prizes. (There is recurring gossip about the reason why there is no Nobel Prize in mathematics. Different versions of the rumor claim that it is due to a rivalry over a woman between Magnus Gösta Mittag-Leffle, the leading Swedish mathematician, and Nobel. The rumor seems to be unjustified.)[31] As of 2018, 590 Nobel Prizes have been awarded to 935 Nobel laureates.

The nomination and selection process

The nomination and selection process for Nobel laureates takes over a year and involves dozens of reviews and expert consultations to determine the winner(s) in each prize category. In all categories but the Nobel Peace Prize, nominations are accepted by invitation only, and nomination forms are sent out in September of the year preceding the awarding of the prize. Invitations for nominations are typically sent to selected professors at universities and to former laureates, who have until January 31 of the following year to submit their nominations. The process typically generates about 250 to 350 unique nominees, and consultation with experts proceeds for several months to assess the worthiness of each candidate. Over the summer, the Nobel Committee writes a report to be submitted in September to the institution responsible for awarding the particular Nobel Prize, and in October a winner is selected through majority vote and announced. Prizes are then awarded in December.

Literature prize: from betting to scandal

In the last 15 years or so, the big sports betting company Ladbrokes has begun taking online bets for the Nobel Prize for literature (only in literature; probably the others attract much less public attention). Bookmakers set fixed odds on all horses in a specific race or on the result of soccer games, but the duty of a Nobel Prize bookmaker is particularly challenging. If you bet on sporting events, you have a lot of data about past achievements, actual injuries, etc. Predictions based on patterns in the existing data are not terribly difficult to make. But for the prize in literature, I am not so sure that it is as easy. You should have a behavioral model of the committee and how they make their decisions.

Betting for the Nobel Prize somehow reflects people's demand, and we should not exclude the possibility that the selection committee totally neglects public opinion. Indeed, we have a long list of unlikely non-winners: Leo Tolstoy, Anton Chekhov, Marcel Proust,

Franz Kafka, Virginia Woolf, James Joyce, and Vladimir Nabokov, among others.

In 2017, Kazuo Ishiguro was awarded the Nobel Prize in literature for his "novels of great emotional force [that have] uncovered the abyss beneath our illusory sense of connection with the world." Margaret Atwood, Ngugi Wa Thiong'o, and Haruki Murakami were given better odds, but still Ishiguro's win was approved by the betting community.

Is there any collusion between the selection committee and the betting process? I would like to bet that there isn't. The committee receives annually about 200 nominations from literary nobilities. Maybe 10 to 15 percent are first-time nominees. I might be wrong, but I believe that first-time nominees initially had a better chance at being selected than they do now. Rabindranath Tagore (1861–1941; winner in 1913), Sinclair Lewis (1885–1951; winner in 1930), Pearl Buck (1892–1973; winner in 1938), Bertrand Russell (1872–1970; winner in 1950), and William Faulkner (1897–1962; winner in 1949) were awarded Nobel Prizes in literature after being nominated in one year only. The ugly sex scandal[32] that led to the cancellation of the award in 2018 might have cataclysmic consequences for the Nobel Prize in literature. We shall see.

The illusion and manipulation of objectivity: Is there any gender bias?

As we noted, 935 individuals have received Nobel Prizes since the prize's inception in 1901. Yet only 51 of them have been women (Marie Curie won twice), meaning that less than 6 percent of Nobel Prize winners have been women. In the sciences, the chemist and science writer Magdolna Hargittai notes, that figure is even smaller, with just 19 women receiving the prize in physics, chemistry, or medicine or physiology since its inception. This abysmally low number has led some to suggest that the Nobel Prize is biased against female scientists.

Certainly, there have been female scientists who have been overlooked for Nobel Prizes during the nomination and selection process, but it may be the case that the relatively low number of women who have been awarded the Nobel Prize is attributable more to implicit bias than explicit discrimination. Stereotypes discouraging women from pursuing education and careers in STEM fields have meant that the number of women receiving doctoral degrees in physics, chemistry, and medicine is relatively low in comparison to their male counterparts, and although these numbers are increasing, they remain lower than population statistics would suggest. Take physics as an example: according to the American Institute of Physics, in 1975 women earned just 5 percent of PhDs in physics; in 2017 the proportion was 18 percent. If less than one in five physics PhDs is awarded to a woman, the odds of being awarded a Nobel Prize in physics is significantly smaller for a woman. Even after PhDs are awarded, implicit bias manifests itself in the barriers that women face in hiring and publication, as studies show that women are more likely to be judged on superficial qualities like appearance and personal information than the quality of their scholarship. In terms of publication, women are less likely to be cited than men and their research is more likely to be attributed to men, and they are underrepresented in journal editorships. Both of these factors work against the likelihood that women will be invited to speak at conferences to present their findings or to be nominated for awards. Since the Nobel selection committees refuse to publish information about nominees until 50 years after the nomination has been submitted, it is difficult to assess the rate at which women are nominated for Nobel Prizes in STEM fields, but based on the state of gender representation in physics, chemistry, and medicine, there is a good chance that women are not being nominated at the same rate as men are. If we acknowledge the kinds of institutionalized biases that greet women at every step of their careers, it is less surprising that so few women have received Nobel Prizes in STEM fields. Although bias certainly exists, the nature of implicit bias is such that it is nearly impossible to pinpoint at which stage in a woman's career her chances of receiving a Nobel Prize were diminished, and it may be the case that the bottleneck occurs well before names ever reach the Nobel Prize committees.

Are the Oscars racist?

Recent debates about the endemic racism of the Academy Awards illustrates that the selection of the "best" reflects the dominant attitudes of the mainstream society, and once again Woolf-Boolf, the self-appointed judge, is with us. Complaints similar to those concerning the Nobel Prize have been launched at the Academy Awards (or Oscars), this time with regard to race. Analysis of the racial composition of Oscar recipients shows a disproportionate overrepresentation for White actors and actresses. Most strikingly, Hispanics make up approximately 16 percent of the US population according to 2010 census data, but they account for just 3 percent of Oscar winners. Furthermore, accusations of racism have surfaced concerning the roles that people of color are awarded in the film industry—it's not just that actors of color are routinely passed over for leading roles, but when they do occupy leading roles, they are often cast in roles that reflect common racial stereotypes like, for example, the maid or mammy, a role played by Viola Davis in the 2011 film *The Help*. One site's analysis of the Oscars received by people of color suggests that over 50 percent of awards went to actors and actresses of color in stereotyped roles. Again, these numbers likely better reflect the effects of long-term systemic biases than the explicit prejudices of specific individuals, although those too have undoubtedly played a role in the whitewashing of Hollywood. But Hollywood itself is often thought of as an old boys' club, where white male executives wield power over which scripts to approve, which films to fund, and which actors to hire. Like the Nobel Prize situation, this means that people of color are likely the victims of implicit bias at every step of the journey, from auditions to award ceremonies. The fact that such bias is implicit rather than explicit makes it no less pernicious, but it does illuminate the fact that reputation, even at the highest echelons of power, is often characterized by such biases. To the extent that we are aware of the ways in which supposedly objective markers of status carry subjective biases, we can better combat them.

The dark side of a success story: the search engine manipulation effect and its possible impact

The color of the hat of your reputation manager

A big industry has emerged with the goal of making websites more visible, and there are search engine optimization companies to perform the job. Even reputation management companies are subject to ranking. As in Western movies, there are characters with white hats and with black hats. There are heroes and villains. Some search engine optimizers, referred to as *ethical hackers*, wear the white hat, but others manipulate information and wear a black hat. Black-hat optimizers attempt to "game" search engine algorithms. As always, in democratic societies, first the community promulgates rules. Then, some people try to evade these rules. We cannot do anything but attempt to identify and neutralize the effects of these troublemakers. Here is a warning you may find useful: a black-hat optimizer can take you to the top of a website ranking in a very short period of time. But strictly speaking, it is totally illegal. If you don't want to get penalized and kill your Google ranking forever, it is strongly recommended that you avoid black-hat optimizers.

Is there a "best online reputation management service"? Who leads this list? Maybe those who really offer their best service. Or maybe those who are the best at managing (not only their own) reputation.

Suggestion 1: You need to deal with your digital reputation.
Suggestion 2: Don't manipulate!

We should live with even completely unfair negative reviews. (Somehow I have never received any unfair positive reviews, each of them was well deserved. How about you?) The fact is that it is possible to learn from every negative review, even those given unfairly. First, read them carefully and cool off! Second, answer quickly and professionally! Rapid action might help put out the fire of negative opinions and help to minimize your ranking loss.

Manipulating the political bias

Manipulating political biases in search results has the potential to impact the voting preferences of undecided voters by 20 percent or more.[33,34] Biasing search rankings constitutes a new type of social influence, and it is occurring on an unprecedented scale. One experiment has suggested that flashing "VOTE" ads to 61 million Facebook users caused more than 340,000 people to vote that day who otherwise would not have done so.

As a continuation of this line of research, Robert Epstein and others have shown that modifying the design of search engines to include alerts about the potential bias of search results can significantly mitigate the aforementioned search engine manipulation effect.

From conformation bias to the propagation of fake news

There are now famous stories about how right-wing hate sites have managed to trick algorithms into associating concepts that require Google's human editors to intervene. One British journal,

The Guardian, published an article in the "Before Trump Era" titled "Google, democracy, and the truth about Internet search." It would be good to find reliable data regarding whether or not right-wing manipulations significantly outnumber those coming from the left. I am not totally sure. Conformation bias is neutral—*everybody* likes to share information that makes us less uncertain. We are ready to forward or share information that connects with other people who think similarly to us. We are quick to pass on even false and problematic content without checking.

We all know now that there are also serious concerns about the effect that search engine manipulation might have on the outcomes of elections. In democratic societies, people use search engines, which are the product of private companies, to research candidates. Even a neutral search engine might influence the outcome of a close election. It is a difficult to answer the question of who has the responsibility for controlling the results. I don't see a better path than trusting in the wisdom of crowds. Hope it will work!

Lessons learned: reputation management as a tool of the ranking game

Reputation is a key element of the ranking game. Reputation can be measured, and there are various strategies for managing our reputations. The strategy one adopts for finding harmony between the struggle for reputation or recognition of external success and the desire for inner peace or internal motivation is a personal decision. Scientists and artists are particularly subject to the reputation game, and the rules for these players are better elaborated than the rules for other communities. Reputation leads to success, and nowadays quantitative methods have emerged to predict scientific and artistic success. Your success is a collective phenomenon, dependent on the opinion of many other people in your community. It was in my mind as I am wrote this sentence.

In any case, our individual wisdom is challenged each day, as algorithms increasingly recommend products, activities, and experiences. The next chapter discusses the nuts and bolts of recommendation systems and gives some advice for coping with them.

8

Inspired by your wish list

How (not to) buy a new lawnmower

The recommendation game: all we need is trust

I would feel confident betting that you have not made any purchasing decision recently without being influenced by the opinion of the Web. As I open Amazon, I see a holiday toy list with a Star Wars Droid Inventor Kit at the top. I consulted TripAdvisor when I returned to Liverpool after decades and needed to find a small hotel near the university, where I actually gave my first talk with the same title that this book has. I don't really use Yelp, since I have my favorite restaurants in Budapest, from Spinoza to Pozsonyi Kisvendéglő. For those of you who live in Manhattan, do you need recommendations? I am not sure, but they exist anyway.

The recommendation game emerged as electronic commerce became part of our everyday lives. There are two types of players, the sellers and the buyers of a set of products, and there is a mechanism to match them using online services. There are games where matching is a better expression than buying and selling: I am now not involved directly in the dating business, but I know that Match.com leads the dating websites, and Jdate is 15th now in popularity. Historically, merchants often knew their potential buyers, and they could therefore make recommendations based on their prior acquisitions. But we don't live the lives of our ancestors, even those just a generation or two ago, and the loss of personal relationships with many sellers has been overcompensated with the availability of many options. Machine learning experts promise to replace personal connections at least partially. The aim of the sophisticated recommendation algorithm is to understand and predict the personal behavior of the consumer. In principle, recommendation algorithms are neutral, and they should not favor the interests of either the sellers or buyers.

Roughly speaking, recommendation systems suggest products to potential consumers based on a number of different reasons. A seller has two different problems. First, here is a new product, say a lawnmower. The goal is now to identify potential buyers. But how? You bought a mower last year. Do you need another one for next year? No, no, no, not at all! But you should have a garden, so you might need some gardening hand tools, right? Second, here is a user, say someone named Liz. So, what are the top three items to recommend to her? We don't live in the world of stereotypes, so my Liz will be recommended some exciting items from the automotive parts and accessories category, followed by the video game *League of Legends*. Finally, there is one more thing she should buy, if she really likes her pet: the Dog DNA Test by Embark!

This procedure is called the *ranking formulation* part of the recommendation problem. High-quality recommendations generated by such systems can transform the user experience from annoying to felicitous, while also contributing to long-term consumer trust and loyalty. Modern recommendation systems combine several strategies by nudging users to specify preferences like these:

- Show me stuff that *my friends like* (collaborative filtering)
- Show me stuff that *I liked in the past* (content-based filtering)
- Show me stuff that *fits my needs* (knowledge-based recommendation)

There are some numbers based on hearsay, such as "35 percent of Amazon's sales come from recommendations." It looks as though our attention is attracted by persuasion techniques. Even if we don't act immediately, we evaluate the slogans we see: "You may also like," "Frequently bought together," "Customers who bought this also bought," and "Recommended for you." Remember, we discussed in Chapter 4 *the paradox of choice*. It is profitable to avoid providing too many choices of products in the same category; an overly long list can simply overwhelm consumers. Another technique helps to make the transition from click to purchase: early algorithms gave recommendations based on what a customer had previously purchased, but real-time recommenders don't need historical data and instead analyze the actual clicking patterns of customers. They focus on the categories consumers are browsing in, the banners and ads they are attracted to, etc., so first-time visitors are immediately grabbed.

In 2017, e-commerce sales accounted for 10 percent of all retail sales worldwide, and the importance of recommendation systems cannot be overestimated. Research has shown that trust is a main factor in the success of a recommender. As buyers, we should feel that recommendations are useful and that the recommendation process is transparent. A growing body of literature analyzes how recommendation systems should demonstrate their usefulness and the transparency of the process, but there is much less advice concerning how to cope with the flood of recommendations.

As I ask around, I think my baby boomer buddies mostly use recommendations to satisfy their immediate needs (to find hotels, to make vacation plans, to buy toys for grandchildren, and maybe to identify restaurants in a new city). As for the other side of the age spectrum, Generation Z is not really interested in my advice, and this is fine. Dear Zs, yes, you are the first generation born into a digital world, and we know that you live online. Here is an unranked list of features that supposedly characterize your attitudes as consumers:[1]

- Use customized smartphone applications released by e-retailers
- Insistence on ease of use
- Desire to feel safe
- Desire to temporarily escape the realities they face
- No loyalty to brands.

So, what's next? The only thing that looks certain is that while more change can be expected, consumers will see more and more ranked lists. There is one more prediction for sure: Generation Z is not the last one, it will be followed by others. I leave for Generation Z to discuss how the Z+ folks will play the recommendation game.

Oh, Netflix

Should I admit that I am not a Netflix subscriber? In my age group only 26 percent of individuals are subscribers, and I am actually thinking that it is not fair to write a book about ranking without having a Netflix subscription, so I may change my mind. To obtain some hands-on experience, I used the subscriptions of youngsters around me. Netflix has realized that it does not have more time than about 90 seconds to grab users' attention by recommending something to watch before they abandon the service and move on to do something else. *Personalized ranking* is the key strategy to ensuring users keep coming back.

A little (not very painful) data science

There are a lot of data regarding our consumption habits being collected not exclusively, but primarily, via social media. In the case of Netflix, the data are specifically about movies and TV shows. There are two types of data: explicit and implicit. When you gave a thumbs-up to *The Post*, your opinion was very explicit. If you watched it twice during a week, this was implicit information about your perception of, mood toward, and relationship with this movie.

In order to make any computational analysis possible, movies are characterized by a number of important extracted features. How "similar" two movies are can be determined by analyzing the similarities between features. As Xavier Amarian, who served as research director for Netflix, writes:

> We know what you played, searched for, or rated, as well as the time, date, and device. We even track user interactions such as browsing or scrolling behavior. All that data is fed into several algorithms, each optimized for a different purpose. In a broad sense, most of our algorithms are based on the assumption that similar viewing patterns represent similar user tastes. We can use the behavior of similar users to infer your preferences.

If you know the distances (i.e., the dissimilarity) between any two items, you can make an ordered list. The smaller the dissimilarity, the better the chance that you will like the recommendation.

A champion algorithm

More precisely, Netflix adopts a family of ranking algorithms, each established for a different purpose. Here is a list of five high-profile algorithms:

- Personalized Video Ranker Algorithm
- Top-N Video Ranker Algorithm
- Trending Now
- Continue Watching
- Video–Video Similarity Algorithm ("sim").

The different algorithms order the entire catalog and prepare ranked lists based on a variety of criteria.[2]

As my software engineer high-school friend advertises in his Skype account: "There are only 10 types of people in the world: Those who understand binary, and those who don't." (For those of you who don't: in binary, you cannot use any digits but 0 and 1, so 2 is not permitted. The numbers you can write are 0, 1, 10, 11, 100, 101, 110, etc. So, in binary, 10 is equal to 2 in our conventional decimal system.) So, for software engineers (speaking now the "decimal language"), only two types of people exist, and the basis of their classification is understanding different number systems. Netflix, by contrast, is personalized into about 2,000 "taste clusters." Who are the members of a cluster? "Not the people who live in the next apartment or house over from you, not the people who live in the same Zip Code, not even the people who live in the same country: the people who tend to enjoy your kind of content."[3] People are clustered based on their viewing habits. The majority of people within a certain cluster will like the same recommendations. Thus it is better to say that recommendations are not personalized, but "clusterized."

The other side of the story: Netflix addiction

Binge watching (the word of the year of 2015) refers to when viewers attempt to watch multiple episodes of a television series in rapid succession. While it shows some correlation with depression and loneliness in viewers, we more or less understand how our brains force us to act as addicts. Episodes of a TV series typically end with some exciting scenes. The trigger is pulled, but we don't see the next picture. These kinds of *cliffhangers* increase a stress-related hormone, so when you push the play button, you see the next episode, and the cycle continues. After several hours of binge watching, you may have a feeling of happiness: "Oh, I spent the whole day here, so it is a big achievement!" We know the underlying neurochemistry: your brain releases dopamine, a reward- and pleasure-related substance, which serves as a reinforcement signal that creates a self-amplifying loop. Still, I am not yet prepared to jump on the bandwagon and to join the "pajamas all day" movement to watch *Black Mirror*.

Fake reviews: they happen but can be filtered

The Bellgrove case

In the travel industry, TripAdvisor is a leader. It's the yardstick for reviews and comparisons of hotels and excursions. Everybody knows anecdotal evidence about major problems with the service. Every system can be gamed, and there is wide discussion about how to cope with the huge problem induced by fake reviews.

One of the most famous TripAdvisor incidents is related to Bellgrove Hotel in Glasgow, Scotland, and it was more the product of a joke than of the intent to generate a fake review. The hostel served about 150 mostly homeless, unemployed men, some of whom had drug and alcohol problems, so it did not have a wonderful reputation. In 2013 a number of jokers gave it a five-star rating, and at a certain point Bellgrove made the top 100 of TripAdvisor's best places to stay! I think TripAdvisor reacted properly: "As this property is a homeless shelter, and therefore doesn't meet our listing guidelines, the listing itself is being removed from TripAdvisor." But the case generated another wave of news, as even the Scottish parliament discussed the hostel's conditions: "that generously, could be considered unsuitable, and, less generously, grim, Dickensian, like a Soviet gulag or similar descriptions."[4]

Benevolence, hoax, permanently beta

Any system established to help people and make a fair profit can be gamed. It was in the news in 2015 that a nonexistent Italian restaurant made the top of TripAdvisor's rankings in a northern Italian small town called Moniga del Garda. *Italia a Tavola*, a leading online newspaper for information on food and wine, made the hoax to prove the manipulability of rankings on the portal. First, a profile was created for an imaginary restaurant, named La Scaletta. Second, fake reviews were generated by a number of conspirators. After receiving excellent reviews, the restaurant made the top spot in the list of the town's restaurants. I am inclined to agree with those who felt the newspaper's methods were unethical. It is more efficient to work directly for a better world than to try to pinpoint the negative side of everything that exists.

We should understand and accept that different types of software are in the *permanent beta* state. "Beta" originally referred to the final stage of software development immediately before the product was launched to the market, and a community of "beta users" gave feedback at this stage of the process. Nowadays, many products remain in this stage and are the subject of continuous improvement.

Maybe all you (we) need is love

One in five relationships now begins online, and nowadays online dating has become an acceptable way to meet people. One of the most successful services, eHarmony, advertises its success stories by defining a bunch of typical categories: "Nearly Gave Up," "Singles with Children," "Re-Connections," "Long Distance," "50+," "Multiple Successes in Family," "So Close Yet So Far," "International," and many more. I have read a number of success stories,[5] most of them banal, something like "We had our first date on December 8, 2010; Bill proposed to me on December 8, 2011;

and we were married on December 8, 2012. From the beginning we had so much in common."

Do the algorithms attempt to identify the banal "so much in common"? In principle, algorithms match users based on their compatibility. Algorithms identify compatibility with some similarity measures. Generally compatibility means that features like beliefs, values, and education are generally similar. However, how often do we have a second group suggesting that "opposites attract"? People from the first group emphasize the importance of cultural similarities, but the others explain their appreciation for their spouse in terms of cultural complements.

Are matching algorithms gender-neutral?

Some youngsters around me claim that the popular social app Tinder has an algorithm that is male-biased. Tinder is an online dating app that matches couples based on one feature only: their physical attraction to one another. You see the picture of a person and decide whether or not you like the look of her. If "yes," your picture is offered to that person. If you mutually like each other, then you two are found to be compatible, and Tinder allows you to initiate conversations. Recent data suggest[6] that Tinder's ranking algorithm tends to disadvantage men, and the assumption that it is male-biased is not justified.

Can algorithms find your #1 romantic partner?

Both my common sense and everyday experience suggest that while you are searching in the database, you shouldn't weep because at least you are actively involved in managing your life—and please remember what was discussed in Chapter 4: "You have the highest chance of finding Ms. or Mr. Right if you date and reject the first 37 percent of your potential mates. The rule has a second part: pick the next person who is better than anyone you have ever dated earlier." I'm not sure how to sell this to your prospective spouse. It doesn't sound like an irresistible proposal: "Honey, I've already used 37 percent of my chances, and you seem to be somewhat better than the horrible chicks I met previously, so my math professor recommended that I marry you!" Still, the pragmatic mind inside me suggests this to you: marry Mr./Ms. Goodenough!

Lessons learned: cautious optimism

Recommendation systems are ubiquitous in our lives. It is difficult to make any purchase without being somehow influenced by large e-commerce systems. Recommendation systems are key elements of any e-commerce system. Nobody can force us to use them, but we do if we trust them. While any such system can be gamed, and some illustrative examples of gaming were given, fake reviews and other tricks can be filtered, and recommendation systems can help us make better choices.

9

Epilogue

Rules of the ranking game—where are we now?

The reality, illusion, and manipulation of objectivity

Like it or not, ranking is with us. It is not a magic bullet that produces order out of chaos, but it is not the product of some random procedure (Figure 9.1). Like it or not, parents and students will carefully study the college ranking lists. If it becomes the accepted view that generally it is not a good idea to make a final decision based only on a formal list, college (and I think many other) ranking systems will serve their purpose: to give some (some!) condensed, often numerical, information. But as in the case of the Hungarian sports newspaper—"Let the objective numbers speak!"—objectivity, more often than not, is an illusion. My advice to students and parents is to make a personalized ranking. Nobody but you can know what factors are important for you. (As the director of a study-abroad program in Budapest, I once overheard this comment from a student: "I prefer to remain in this dorm, which has an excellent Internet connection—even if there are some bugs in the building.") Human ranking is subject to cognitive bias, and computational ranking procedures are based on databases and algorithms. Databases and algorithms are generally biased, but they are not totally random and they do reflect some aspects of objectivity.

If you don't like the rank that you or your organization received, after allowing yourself five minutes of irritation, it is a good idea

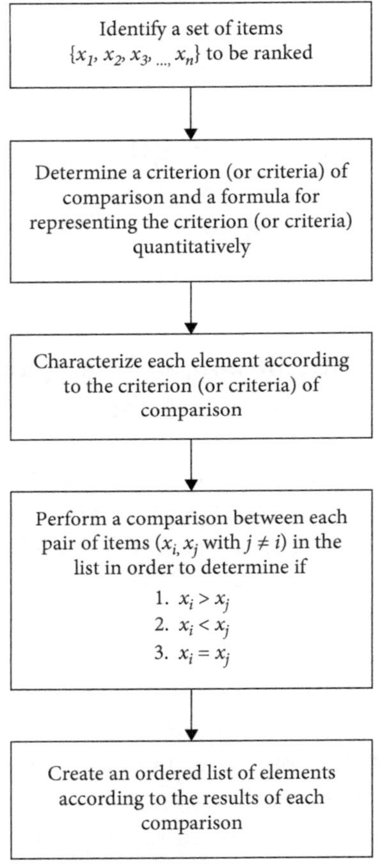

Fig. 9.1 A flowchart to describe the process of generating a ranked list.

to consider whether (1) the ranker is just a malignant evildoer or (2) the review might actually contain some elements of truth and there is a chance to improve your performance. I might be too idealistic, my advice is this: "cool off and think." As concerns manipulation, I do believe that in the long term nothing else matters but the opinion of your community. I try to believe that the wisdom of the crowd exceeds the madness of the crowd. This is not so easy, as a famous historical example shows.

The South Sea Bubble occurred in England between 1711 and 1722. The British government offered a deal to the South Sea Company to finance a significant state debt that had emerged during the War of Spanish Succession. The South Sea Company traded with South America (excluding Brazil, as it was a Portuguese territory). After a rumor that the South Sea Company had been granted full use of Latin American ports, "It became extremely fashionable to own South Sea Company shares."[1] It turned out at a certain point that the actual commerce did not produce profit for company leaders, and money was generated mostly from issuing stocks, so their shares became strongly overvalued. Soon after the owners started to sell shares, there was a panic among shareholders and the market crashed. Factors like speculation, unrealistic expectations, and corruption contributed to the emergence of the bubble. Sir Isaac Newton (1643–1727), who was a scientist, the master of the mint, and a certifiably rational man, first saw the bubble but later lost a lot of money because of it. He sold his 7,000 pounds worth of stock in April for a profit of 100 percent, but something induced him to reenter the market at the top, and he lost £20,000, leading him to declare, "I can calculate the movement of the stars, but NOT the madness of men." Jonathan Swift (1667–1745) also lost a large amount of money, which motivated him to write a satire about British society (*Gulliver's Travels*) and "The Bubble: A Poem":

> The Nation too, too late will find
> Computing all their Cost and Trouble
> Directors Promises but Wind
> South Sea at best a mighty Bubble.

I think Swift would write a fantastic bestseller about Brexit.

In any case, as President Abraham Lincoln famously stated: "You can fool all the people some of the time and some of the people all the time, but you cannot fool all the people all the time."

Comparison is very human

It's human nature to compare ourselves to others. The question is how to cope with the results of these comparisons. At the end of this book you may ask: "Okay, this guy wrote a book about ranking. How does he himself play the ranking game and how does he compare himself to his fellows?" Let me be very personal: I have a decent h-index, but it is still lower than that of many of my peers, and due to the nature of this index, it cannot be changed at this stage of my career. I know that my relatively low h-index is the price I paid for some of the decisions I made. First, my first job was in industry (at the computer center of the Danube Oil Company, an oil refinery that processed crude oil from the Soviet Union), not in academia, so I made a slow start. (It is impossible for American minds to understand the surreal career paths we had in Hungary in the 1970s.) Second, my mathematician friend (the one who never had a car) and I discovered/constructed an algorithm (for stochastic simulation of chemical reactions), but we did not have the ideas and resources to publish properly. The same algorithm was published by an American scientist who worked for the Naval Weapons Center in China Lake, California, maybe a year later, and the papers generated 20,000-plus citations. (Sour grapes, I know.) Third, instead of concentrating on one specific field and method, I have been working on different topics, from chemistry via neuroscience to political science to patent citation analysis. Fourth, I have spent years writing monographs for books in Hungarian instead of working on scientometrically more efficient papers. Fifth, I accepted a very prestigious professorship at a liberal arts college, and I spend only a few months annually with my Hungarian research group. I get my salary for teaching, which I like, but I still have had less time for research.

Am I envious of a peer who has 10 times more citations and an h-index four times higher than my own? Yes! But still, I can write books, and I enjoyed very much the two years I invested in writing this text. Is it a success? Nobody knows whether or not a book will be a bestseller. I believe I combine well downward and upward comparisons to accept my place in the scientific horse race and preserve my enthusiasm to create something new.

Lists help us comprehend incoming information

While we are not very good at memorizing long lists, lists still help us process sensory information. In addition, reading listicles can give us the impression that we have gained complete knowledge about a certain topic. Lists also help to organize our daily activity by allowing us to decide on the relative importance of the different projects involved.

Social ranking has an evolutionary root

Dominance hierarchies are very efficient structures at very different levels of evolution. They have a major role in reducing conflict and maintaining social stability. Dominance is based on aggression and manipulation, and it frequently serves the self-interest of the dominant leader. Prestige, a different mechanism for helping people outrank others, is based on knowledge, and it generally serves the interest of a community. While there are elements of network organization in our society, we have some concern about how to treat the return in recent years of hierarchical authoritarianism.

We humans constructed ranking algorithms

One of the most important tasks of any society is to make collective decisions based on individual opinions. There are many voting systems. Although none of them is perfect, they are generally better than leaving the decision to a subjective or manipulative individual. Some ranking procedures in moral, religious, and legal systems lead to nontransitive cycles. We also know that the result of the PageRank algorithm depends on the numerical value of a parameter, and rank reversal (i.e., a change in the rank ordering) may happen.

Ranking games are with us; even hermits cannot avoid playing. Whether you are a job applicant or a member of a search committee, you will have to engage with scoring and ranking: either you'll be scored and ranked or you'll be the one doing the scoring and ranking. Like it or not, there is an increased demand for transparency, accountability, and comparability of institutions and individuals. Metrics are more useful than totally subjective evaluations. Can metrics be gamed? Certainly, and Campbell's law teaches us about the illusion of objectivity. Still, when you see a list of colleges or countries based on any criteria, nobody will tell you that the ranked list is produced by some random algorithms. Ranked lists of colleges and countries are not worse than the educated guesses of experts. I suggest the rule "trust, but with caution."

Balance reputation, external success, and internal peace

One of the Founding Fathers of the United States, Thomas Paine (1737–1809), a political theorist and activist, said: "Reputation is what men and women think of us; character is what God and angels know of us." Artists and scientists are more reputation-driven than many other people. Digital reputation matters, since we more or less trust what we read on the Internet. While our obsession with metrics has generated a huge industry, even in our success-oriented society the best strategy is to try to keep a balance between the struggle for reputation and external success and the desire for internal peace.

Recommendation systems help us to think about our options

Every day we rank many of our options by using recommendation systems. The success of these systems depends on our trust. We all know that fake reviews occur, but I think the combination of human and computational intelligence can filter them out. Again, I suggest the rule "trust, but with caution!"

Controlling the web: Who has the last word, the human or the computer?

Computer scientists design ranking algorithms, and of course, computers can now process huge datasets using these algorithms. As we have seen, we are not always happy with the results, so we might ask whether, when, and how the results of a ranking algorithm should be controlled by *content curators*. Classically, museums have curators to select artworks for display in a specific exhibition. Whether or not we should control the results of our algorithms, and if so, how rankings produced by soulless algorithms should be modified, will be a battlefield in the coming decades.

Cathy O'Neil, a mathematician and blogger, argues in her book *Weapons of Math Destruction*[2] that Google will ultimately have to hire human editors. She might be right, and hopefully these human editors would come to power with real knowledge and prestige and not by manipulation and dominance. The age-old question "What came first: the chicken or the egg?" is still with us. Now we have a new question, too: "Who has the last word: the human or the computer?" I leave the question and the possible answers to Generations Z and Z+.

Notes

Chapter 1

1. If you are not familiar with such an image, take a moment to Google "cat lion mirror," and you will not be disappointed.

Chapter 2

1. Adam Galinsky and Maurice Schweitzer, *Friend & Foe: When to Cooperate, When to Compete, and How to Succeed at Both* (New York: Penguin Random House, 2015).
2. Alfie Kohn, *Punished by Rewards: The Trouble with Gold Stars, Incentive Plans, A's, Praise, and Other Bribes* (Boston: Houghton Mifflin, 1999).
3. Haleh Yazdi, "Intrinsically motivated," Usable Knowledge, September 11, 2016, https://www.gse.harvard.edu/news/uk/16/09/intrinsically-motivated.
4. Robert Fulghum, *All I Really Need to Know I Learned in Kindergarten* (New York: Penguin Random House, 1986).
5. Yi Luo, Simon B. Eickhoff, Sébastien Hétu, and Chunliang Feng, "Social comparison in the brain: a coordinate-based meta-analysis of functional brain imaging studies on the downward and upward comparisons," *Human Brain Mapping* 39, no. 1 (January 2018): pp. 440–458, https://doi.org/10.1002/hbm.23854.
6. "West Germany's 1954 World Cup win may have been drug-fuelled, says study," *The Guardian*, October 27, 2010, https://www.theguardian.com/football/2010/oct/27/west-germany-1954-drugs-study.
7. Rob Hughes, "Doping study throws shadow over Germany's success," *New York Times*, August 6, 2013, https://www.nytimes.com/2013/08/07/sports/soccer/Doping-Study-Throws-Shadow-Over-Germanys-Success.html.
8. Figure 9.1 shows an algorithm for generating a ranked list.

9. Stanley Smith Stevens, "On the theory of scales of measurement," *Science* 103, no. 2684 (June 1946): pp. 677–680, http://science.sciencemag.org/content/103/2684/677.
10. Paul F. Velleman and Leland Wilkinson, "Nominal, ordinal, interval, and ratio typologies are misleading," *American Statistician* 47, no. 1 (1993): pp. 65–72, https://www.tandfonline.com/doi/abs/10.1080/00031305.1993.10475938.
11. Celia Vimont, "Numbers don't tell the whole story: experts say better pain assessment measures needed," *Practical Pain Management*, February 7, 2017, https://www.practicalpainmanagement.com/patient/resources/understanding-pain/numbers-dont-tell-whole-story-experts-say-better-pain.
12. Lewis Richmond, "Emptiness: the most misunderstood word in Buddhism," *Huffington Post*, March 6, 2013, https://www.huffingtonpost.com/lewis-richmond/emptiness-most-misunderstood-word-in-buddhism_b_2769189.html.
13. Robert Kaplan, *The Nothing That Is: A Natural History of Zero* (New York: Oxford University Press, 2000).
14. Amir D. Aczel, *Finding Zero: A Mathematician's Odyssey to Uncover the Origins of Numbers* (New York: Macmillan, 2015).
15. Hannah Devlin, "Much ado about nothing: ancient Indian text contains earliest zero symbol," *The Guardian*, September 13, 2017, https://www.theguardian.com/science/2017/sep/14/much-ado-about-nothing-ancient-indian-text-contains-earliest-zero-symbol.
16. Andreas Nieder, "Representing something out of nothing: the dawning of zero," *Trends in Cognitive Science* 20, no. 11 (November 2016): pp. 830–842, https://www.ncbi.nlm.nih.gov/pubmed/27666660.
17. Karl Popper, *The Self and Its Brain* (Berlin: Springer, 1977).
18. "Yehudah HaNasi (Judah the Prince)," Jewish Virtual Library, accessed February 10, 2019, https://www. jewishvirtuallibrary.org/yehudah-hanasi-judah-the-prince.
19. Bert and Kate McKay, "10 tests, exercises, and games to heighten your senses and situational awareness," *Art of Manliness*, March 15, 2016, https://www.artofmanliness.com/about-2/.
20. Alexander Luria, *The Mind of a Mnemonist: A Little Book About a Vast Memory* (New York: Basic Books, 1968).
21. Josette Akresh-Gonzales, "Spaced repetition: the most effective way to learn," NEJM Knowledge+, November 19, 2015, https://knowledgeplus.nejm.org/blog/spaced-repetition-the-most-effective-way-to-learn/.

22. Marc Augustin, "How to learn effectively in medical school: test yourself, learn actively, and repeat in intervals," *Yale Journal of Biology and Medicine* 87, no. 2 (June 2014): pp. 207–212, https://www.ncbi.nlm. nih.gov/pmc/articles/PMC4031794/#.
23. "Does Anki really work?", Reddit, accessed February 10, 2019, https://www.reddit.com/r/Anki/comments/2w1mgm/does_anki_really_work/.
24. Claudia Hammond, "Nine psychological reasons why we love lists," *BBC*, April 13, 2015, http://www.bbc. com/future/story/20150410-9-reasons-we-love-lists.
25. https://www.youtube.com/watch?v=qMQj7YZ9eOU.
26. James Clear, "The Ivy Lee Method: the daily routine experts recommend for peak productivity," Accessed February 10, 2019, https://jamesclear.com/ivy-lee.
27. "Warren Buffett's 5/25 rule will help you focus on the things that really matter," Constant Renewal, accessed February 10, 2019, https://constantrenewal.com/buffett-5-25-rule/.
28. Bryan D. Jones and Frank R. Baumgartner, *The Politics of Attention: How Government Prioritizes Problems* (Chicago: University of Chicago Press, 2005).
29. Julie Compton, "Forget to-do lists: use a might-do list to work smarter," *NBC News*, April 11, 2017, https://www.nbcnews.com/better/careers/do-lists-don-t-work-use-might-do-list-work-n744831.
30. Brent DiCrescenzo and Adam Selzer, "The 17 most notorious mobsters from Chicago," TimeOut, March 4, 2015, https://www.timeout.com/chicago/things-to-do/the-17-most-notorious-mobsters-from-chicago.
31. James Surowiecki, *The Wisdom of Crowds: Why the Many Are Smarter Than the Few and How Collective Wisdom Shapes Business, Economies, Societies and Nations* (New York: Doubleday, 2004)
32. Jan Lorenz, Heiko Rauhut, Frank Schweitzer, and Dirk Helbing, "How social influence can undermine the wisdom of crowd effect," *Proceedings of the National Academy of Sciences of the United States of America* 108, no. 22 (May 2011): pp. 9020–9025, https://doi.org/10.1073/pnas.1008636108.
33. Scott E. Page, *The Difference: How the Power of Diversity Creates Better Groups, Firms, Schools, and Societies* (Princeton: Princeton University Press, 2007).
34. Mirta Galesic, Daniel Barkoczi, and Konstantinos Katsikopoulos, "Smaller crowds outperform larger crowds and individuals in realistic task conditions," *Decision* 5, no. 1 (January 2018): pp. 1–15, http://dx.doi. org/10.1037/dec0000059.

Chapter 3

1. https://amboselibaboons.nd.edu/.
2. Eric Bonabeau, Guy Theraulaz, and Jean-Louis Deneubourg, "Dominance orders in animal societies: the self-organization hypothesis revisited," *Bulletin of Mathematical Biology* 61, no. 4 (July 1999): pp. 727–757, https://doi.org/10.1006/bulm.1999.0108.
3. Mathias Franz, Emily McLean, Jenny Tung, Jeanne Altmann, and Susan C. Alberts, "Self-organizing dominance hierarchies in a wild primate population," *Proceedings of the Royal Society B: Biological Sciences* 282, no. 1814 (September 2014), https://doi.org/10.1098/rspb.2015.1512.
4. Elizabeth Hobson and Simon DeDeo, "Social feedback and the emergence of rank in animal society," *PLoS Computational Biology* 11, no. 9 (September 2015), https://doi.org/10.1371/journal.pcbi.1004411.
5. Jon Maner, "Dominance and prestige: a tale of two hierarchies," *Current Directions in Psychological Science* 26, no. 6 (November 2017): pp. 526–531, https://doi.org/10.1177/0963721417714323.
6. Hemant Kakkar and Niro Sivanathan, "When the appeal of a dominant leader is greater than a prestige leader," *Proceedings of the National Academy of Sciences of the United States of America* 114, no. 26 (June 2017): pp. 6734–6739, https://doi.org/10.1073/pnas.1617711114.
7. Niro Sivanathan and Hemant Kakkar, "Explaining the global rise of 'dominance' leadership," *Scientific American*, November 14, 2017, https://www.scientificamerican.com/article/explaining-the-global-rise-of-ldquo-dominance-rdquo-leadership/.
8. Edward O. Wilson, *Sociobiology: The New Synthesis* (Cambridge, MA: Harvard University Press, 1975).
9. Jerome H. Barkow, Leda Cosmides, and John Tooby, eds., *The Adapted Mind: Evolutionary Psychology and the Generation of Culture* (New York: Oxford University Press, 1992).
10. "8 reasons a little adrenaline can be a very good thing," Mental Floss, accessed February 10, 2019, http://mentalfloss.com/article/71144/8-reasons-little-adrenaline-can-be-very-good-thing.
11. Eric A. Smith, "Why do good hunts have higher reproductive success?," *Human Nature* 15, no. 4 (December 2004): pp. 342–363, https://link.springer.com/article/10.1007 2Fs12110-004-1013-9.
12. Martin A. Nowak and Karl Sigmund, "Evolution of indirect reciprocity," *Nature* 437, (October 2005): pp. 1291–1298, https://doi.org/10.1038/nature04131.

13. Anna Zafeiris and Tamás Vicsek, *Why We Live in Hierarchies? A Quantitative Treatise* (Berlin: Springer, 2018).
14. Peter Turchin and Sergey Gavrilets, "Evolution of complex hierarchical societies," *Social Evolution and History* 8, no. 2 (September 2009): pp. 167–198, http://www.socionauki.ru/journal/files/seh/2009_2/ evolution_of_complex_hierarchical_societies.pdf.
15. Jung-Kyoo Choi and Samuel Bowles, "The coevolution of parochial altruism and war," *Science* 318, no. 5850 (October 2007): pp. 636–640, http://science.sciencemag.org/content/318/5850/636.
16. Thank you to Bryan D. Jones for making this point clear.
17. "Aztec social structure," Jamail Center for Legal Research at Texas Law, accessed February 10, 2019, http://tarlton.law.utexas.edu/aztec-and-maya-law/aztec-social-structure.
18. "Pyramid of feudal hierarchy," Hierarchy Structure, accessed February 10, 2019, https://www.hierarchystructure.com/pyramid-of-feudal-hierarchy/.
19. "Toga-ther, we will rule history!", Hello World Civ, April 12, 2016, https://helloworldciv.squarespace.com/blog/toga-ther-we-will-rule-history.
20. Mami Suzuki, "Bowing in Japan: everything you've ever wanted to know about how to bow, and how not to bow, in Japan," Tofogu, https://www.tofugu.com/japan/bowing-in-japan/.
21. Arnold K. Ho, Jim Sidanius, Nour Kteily, Jennifer Sheehy-Skeffington, Felicia Pratto, Kristin E. Henkel, Rob Foels, and Andrew L. Stewart, "The nature of social dominance orientation: theorizing and measuring preferences for intergroup inequality using the new SDO7 scale," *Journal of Personality and Social Psychology* 109, no. 6 (December 2015): pp. 1003–1028, https://www.ncbi.nlm.nih.gov/pubmed/26479362.
22. Caroline F. Zink, Yunxia Tong, Qiang Chen, Danielle S. Bassett, Jason L. Stein, and Andreas Meyer-Lindenberg, "Know your place: neural processing of social hierarchy in humans," *Neuron* 58, no. 2 (April 2008): pp. 273–283, https://www.ncbi.nlm.nih.gov/pubmed/18439411.
23. Dharshan Kumaran, Andrea Banino, Charles Blundell, Demis Hassabis, and Peter Dayan, "Computations underlying social hierarchy learning: distinct neural mechanisms for updating and representing self-relevant information," *Neuron* 92, no. 5 (December 2016): pp. 1135–1147, https://www.ncbi.nlm.nih.gov/pmc/ articles/PMC5158095/.
24. Bebhinn Donnelly-Lazarov, *Neurolaw and Responsibility for Action: Concepts, Crimes, and Courts* (Cambridge, UK: Cambridge University Press, 2018).

25. Barry Wellman, "Physical place and cyberplace: the rise of personalized networking," *International Journal of Urban and Regional Research* 25, no. 2 (June 2001): pp. 227–252, https://doi.org/10.1111/1468-2427.00309.
26. Manuel Castells, *The Rise of The Network Society* (Hoboken, NJ: Wiley, 2000).
27. Niall Ferguson, *The Square and the Tower: Networks and Power, from the Freemasons to Facebook* (New York: Penguin Random House, 2018).
28. Steven Levitsky and Daniel Ziblatt, *How Democracies Die* (New York: Penguin Random House, 2018).
29. David Van Reybrouck, *Against Elections: The Case for Democracy* (New York: Penguin Random House, 2017).
30. Joseph A. Califano, Jr., *Our Damaged Democracy: We the People Must Act* (New York: Simon and Schuster, 2018).
31. Yascha Mounk, *The People vs. Democracy: Why Our Freedom Is in Danger and How to Save It* (Cambridge, MA: Harvard University Press, 2018).
32. "Current partisan gerrymandering cases," Brennan Center for Justice, April 26, 2017, https://www.brennancenter.org/analysis/ongoing-partisan-gerrymandering-cases.

Chapter 4

1. "10 tallest buildings in the world," World Atlas, February 10, 2019, https://www.worldatlas.com/articles/10-tallest-buildings-in-the-world.html.
2. "The most influential people of all time," Ranker, accessed February 10, 2019, https://www.ranker.com/ crowdranked-list/the-most-influential-people-of-all-time.
3. Amy Langville and Carl Meyer, *Who's #1?: The Science of Rating and Ranking* (Princeton, NJ: Princeton University Press, 2012).
4. Andrzej Wierzbicki, "The problem of objective ranking: foundations, approaches and applications," *Journal of Telecommunications and Information Technology*, no. 3 (March 2008): pp. 15–23, https://pdfs.semanticscholar.org/4968/788d7bb8570f92c6390fab1ef673f127a500.pdf.
5. Kenneth Arrow, *Social Choice and Individual Values* (Hoboken, NJ: Wiley, 1951).
6. Milton Friedman, *Essays in Positive Economics* (Chicago: University of Chicago Press, 1953).
7. John Rawls, *A Theory of Justice* (Oxford: Oxford University Press, 1972).
8. Amartya Sen, *Choice, Welfare, and Measurement* (Cambridge, MA: Harvard University Press, 1997).

9. Brian Christian and Tom Griffiths, *Algorithms to Live By: The Computer Science of Human Decisions* (New York: HarperCollins, 2016).
10. Daniel Kahneman and Amos Tversky, "Prospect theory: an analysis of decision under risk," *Econometrica* 47, no. 2 (March 1979): pp. 263–292, https://www.jstor.org/stable/1914185.
11. Dan Ariely, *Predictably Irrational: The Hidden Forces That Shape Our Decisions* (New York: HarperCollins, 2008).
12. Klaus Mathis and Ariel David Steffen, "From rational choice to behavioural economics: theoretical foundations, empirical findings and legal implications," in *European Perspectives on Behavioural Law and Economics*, edited by Klaus Mathis (Berlin: Springer, 2016).
13. Eliza Thompson, "13 shark movies that will make you avoid the water forever," April 10, 2018, https://www.cosmopolitan.com/entertainment/movies/a9605910/best-shark-movies/.
14. "'Busiest political betting day in history': bookmakers bet on Britain staying in EU," *The Telegraph*, June 23, 2016, https://www.telegraph.co.uk/news/2016/06/23/busiest-political-betting-day-in-history-bookmakers-bet-on-brita/.
15. "How bookies blew the Brexit call," MarketWatch, June 24, 2016, https://www.marketwatch.com/story/how-bookies-blew-the-brexit-call-2016-06-24.
16. Sir Francis Bacon, *Novum Organum*, ed. Joseph Devey (New York: P. F. Collier, 1902).
17. Jonas T. Kaplan, Sarah I. Gimbel, and Sam Harris, "Neural correlates of maintaining one's political beliefs in the face of counterevidence," *Scientific Reports Volume* 6, no. 39589 (December 2016), https://doi.org/10. 1038/srep39589.
18. Hilaire Gomer, "Loss aversion bias in economics and decision making," Capital, September 6, 2017, https://capital.com/loss-aversion-bias.
19. Ine Beyens, Eline Frison, and Steven Eggermont, "'I don't want to miss a thing': adolescents' fear of missing out and its relationship to adolescents' social needs, Facebook use, and Facebook-related stress," *Computers in Human Behavior* 64, no. 11 (November 2016): pp. 1–8, https://doi.org/10.1016/j.chb.2016.05.083.
20. Marina Milyavskaya, Mark Saffran, Nora Hope, and Richard Koestner, "Fear of missing out: prevalence, dynamics, and consequences of experiencing FOMO," *Motivation and Emotion* 42, no. 5 (October 2018): pp. 725–737, https://doi.org/10.1007/s11031-018-9683-5.

21. Barry Schwartz, "The tyranny of choice," *Scientific American Mind* (December 2004), https://www.scientificamerican.com/article/the-tyranny-of-choice/.
22. Richard Thaler and Cass Sunstein, *Nudge: Improving Decisions About Health, Wealth, and Happiness* (New Haven, CT: Yale University Press, 2008).
23. Pelle Hansen and Andreas Jespersen, "Nudge and the manipulation of choice: a framework for the responsible use of the nudge approach to behaviour change in public policy," *European Journal of Risk Regulation* 4, no. 1 (January 2013): pp. 3–28, https://doi.org/10.1017/S1867299X00002762.
24. William Gehrlein, *Condorcet's Paradox* (Berlin: Springer, 2006)
25. Marianne Freiberger, "Electoral impossibilities," *+Plus Magazine*, April 9, 2010, https://plus.maths.org/ content/os/latestnews/jan-apr10/election/index.
26. Ibid.
27. John Barrow, "Outer space: how to rig an election", *+Plus Magazine*, March 1, 2008, https://plus.maths.org/content/outer-space-how-rig-election.
28. Justin Wise, "Maine votes to keep ranked-choice voting system," *The Hill*, June 13, 2018, https://thehill.com/homenews/campaign/392045-maine-votes-to-keep-ranked-choice-voting-system.
29. "Rules governing the administration of elections determined by ranked-choice voting," Maine Department of the Secretary of State, accessed February 10, 2019, https://www.maine.gov/sos/cec/elec/upcoming/pdf/250rcvnew.pdf.
30. Michel Balinski and Rida Laraki, *Majority Judgment: Measuring, Ranking, and Electing* (Cambridge, MA: MIT Press, 2010).
31. Ibid.
32. Shlomo Naeh and Uzi Segal, "The Talmud on transitivity," Boston College Working Papers in Economics 687 (Boston College, 2008), https://ideas.repec.org/p/boc/bocoec/687.html.
33. Barak Medina, Shlomo Naeh, and Uzi Segal, "Ranking ranking rules," *Review of Law & Economics* 9, no. 1 (2013): pp. 73–96, https://www.bc.edu/content/dam/files/schools/cas_sites/economics/pdf/Uzi-papers/2013 20rle20medina20naeh20ranking.pdf.
34. Amir Salihefendic, "How Reddit ranking algorithms work," Medium, December 8, 2015, https://medium.com/hacking-and-gonzo/how-reddit-ranking-algorithms-work-ef111e33d0d9.

35. Gourab Ghoshal and Albert-László Barabási, "Ranking stability and super-stable nodes in complex networks," *Nature Communications* 2 no. 394 (July 2011), https://doi.org/10.1038/ncomms1396.
36. Albert Lázló Barabási and Réka Albert, "Emergence of scaling in random networks," *Science* 286, no. 5439 (October 1999): pp. 509–512, http://science.sciencemag.org/content/286/5439/509.

Chapter 5

1. David Dunning, "We are all confident idiots," *Pacific Standard*, October 27, 2014, https://psmag.com/social-justice/confident-idiots-92793.
2. Justin Kruger and David Dunning, "Unskilled and unaware of it: how difficulties in recognizing one's own incompetence lead to inflated self-assessments." *Journal of Personality and Social Psychology* 77, no. 6 (December 1999): pp. 1121–1134, https://www.ncbi.nlm.nih.gov/pubmed/10626367.
3. Elemér Lábos, "A dezinformatika alapvonalai," *Valóság* 37, no. 5 (1999): pp. 46–67 [in Hungarian].
4. https://www.youtube.com/watch?v=BdnH19KsVVc.
5. https://www.youtube.com/watch?v=Q_UvfESHUjI.
6. William Poundstone, "The Dunning–Kruger president," *Psychology Today*, January 21, 2017, https://www. psychologytoday.com/us/blog/head-in-the-cloud/201701/the-dunning-kruger-president.
7. David Brooks, "When the world is led by a child," *New York Times*, May 15, 2017, https://www.nytimes.com/2017/05/15/opinion/trump-classified-data.html.
8. Michael Wolff, *Fire and Fury* (New York: Henry Holt and Company, 2018).
9. Thank you to Slava Osaulenko for his excellent lecture during his visit to Kalamazoo.
10. Alex Altman, "No president has spread fear like Donald Trump," *Time*, February 9, 2017, http://time.com/4665755/donald-trump-fear/.
11. Marc Santora, "Orban campaigns on fear, with Hungary's democracy at stake," *New York Times*, April 7, 2018, https://www.nytimes.com/2018/04/07/world/europe/hungary-viktor-orban-election.html.
12. George W. Bush, "Address to the Joint Session of Congress," September 20, 2001, http://edition.cnn.com/2001/US/09/20/gen.bush.transcript./
13. Andrew Clark, "Murdoch's Wall Street shuffle," *The Guardian*, June 22, 2008, https://www.theguardian.com/media/2008/jun/23/wallstreetjournal.newscorporation.

14. Nemil Dalal, "Today's biggest threat to democracy isn't fake news—it's selective facts," Quartz, November 16, 2017, https://qz.com/1130094/todays-biggest-threat-to-democracy-isnt-fake-news-its-selective-facts/.
15. Robert Ensor, *The Era of Violence*, vol. 12, The New Cambridge Modern History, ed. David Thomson (Cambridge, UK: Cambridge University Press, 1960).
16. George Orwell, *Animal Farm* (London: Secker & Warburg, 1945).
17. Tom Stafford, "How liars create the 'illusion of truth,'" *BBC*, October 26, 2016, http://www.bbc.com/future/story/20161026-how-liars-create-the-illusion-of-truth.
18. "George Clooney talks sustainability at Nespresso," 3BL Media, September 11, 2017, https://3blmedia.com/News/George-Clooney-Talks-Sustainability-Nespresso.
19. Cass Sunstein, "The Future of Free Speech," *The Little Magazine* (March–April 2001), http://www.littlemag.com/mar-apr01/cass.html.
20. Zeynep Tufekçi, "It's the (democracy-poisoning) golden age of free speech," *Wired*, January 16, 2018, https://www.wired.com/story/free-speech-issue-tech-turmoil-new-censorship/.
21. Hannah Arendt, "Truth and politics," *New Yorker*, February 25, 1967, http://www.hannaharendtcenter.org/truth-in-politics-hannah-arendt/.
22. Ibid., 23.
23. Ibid., 24.
24. Cass Sunstein, *Republic.com* 2.0 (Princeton, NJ: Princeton University Press, 2009).
25. "The most manipulative characters in film," Ranker, accessed July 16, 2018, https://www.ranker.com/list/most-manipulative-movie-characters/anncasano.
26. "Who are some of the most manipulative leaders in history?", Quora, accessed February 10, 2019, https://www.quora.com/Who-are-some-of-the-most-manipulative-leaders-in-history.
27. Michael Bratton, Boniface Dulani, and Eldred Masunungure, "Detecting manipulation in authoritarian elections: survey-based methods in Zimbabwe," *Electoral Studies* 42 (June 2016): pp. 10–21, https://doi.org/10.1016/j.electstud.2016.01.006.
28. H. James Harrington, quoted in *CIO Enterprise*, September 15, 1999.
29. Donald Campbell, "Assessing the impact of planned social change," *Evaluation and Program Planning* 2, no. 1 (1979): pp. 67–90, https://doi.org/10.1016/0149-7189(79)90048-X.

30. C. A. E. Goodhart, *Monetary Theory and Practice: The UK Experience* (Berlin: Springer, 1975).
31. Paul Craig Roberts and Katharine LaFollette, *Meltdown: Inside the Soviet Economy* (Washington, DC: The Cato Institute, 1990).
32. Jerry Muller, *The Tyranny of Metrics* (Princeton, NJ: Princeton University Press, 2018).
33. Robert Merton, *The Sociology of Science* (Chicago: University of Chicago Press, 1973).
34. "To him who hath shall be given and from him who hath not, shall be take away even what he hath," from the parable of the three servants.
35. Richard Münch and Len Ole Schäfer, "Rankings, diversity, and the power of renewal in science: a comparison between Germany, the UK, and the US," *European Journal of Education* 49, no. 1 (March 2014): pp. 60–76, https://doi.org/10.1111/ejed.12065.
36. Leanna Garfield, "13 cities that are starting to ban cars," Business Insider, June 1, 2018, https://www.businessinsider.com/cities-going-car-free-ban-2017-8.
37. "The minute you start to measure is the minute it starts to go wrong," mmitII, July 6, 2012, https://mmitii.mattballantine.com/2012/07/06/the-minute-you-start-to-measure-is-the-minute-it-starts-to-go-wrong/.
38. Max Nisen, "Why GE had to kill its annual performance reviews after more than three decades," Quartz, August 13, 2015, https://qz.com/428813/ge-performance-review-strategy-shift/.
39. "Performance development at GE (PD@GE)," Fast Company, accessed February 10, 2019, https://www.fastcompany.com/product/performance-development.
40. Richard Dawkins, *The Selfish Gene* (Oxford: Oxford University Press, 1976).
41. Amy Graff, "Yahoo slapped with lawsuit for gender discrimination against men," SF Gate, February 28, 2018, https://www.sfgate.com/news/article/Yahoo-lawsuit-Marissa-Mayer-discrimination-men-9926263.php/.
42. "History of Equifax, Inc.," Funding Universe, accessed February 10, 2019, http://www.fundinguniverse.com/company-histories/equifax-inc-history/.
43. "What's in my FICO scores," myFICO, accessed February 10, 2019, https://www.myfico.com/credit-education/whats-in-your-credit-score/.
44. Andrew Ferguson, *The Rise of Big Data Policing: Surveillance, Race, and the Future of Law Enforcement* (New York: NYU Press, 2017).

45. Indrė Žliobaitė, "Fairness-aware machine learning: a perspective," arXiv, August 2, 2017, https://arxiv.org/abs/1708.00754v1.
46. Mark Kear, "Playing the credit score game: algorithms, 'positive' data, and the personification of financial objects," *Economy and Society* 46, no. 3-4 (2017): pp. 346–368, https://doi.org/10.1080/03085147.2017.1412642.
47. John von Neumann, "Can we survive technology?," *Fortune* (June 1955), http://fortune.com/2013/01/13/can-we-survive-technology/Ranking games.

Chapter 6

1. Mathew S. Isaac and Robert M. Schindler, "The top-ten effect: consumers' subjective categorization of ranked lists," *Journal of Consumer Research* 40, no. 6 (April 2014): pp. 1181–1202, https://www.jstor.org/ stable/10.1086/674546.
2. Mathew S. Isaac, Aaron R. Brough, and Kent Grayson, "Is top 10 better than top 9? The role of expectations in consumer response to imprecise rank claims," *Journal of Marketing Research* 53, no. 3 (June 2016): pp. 338–353, https://doi.org/10.1509/jmr.14.0379.
3. Carl Kořistka, Der höhere polytechnische Unterricht in Deutschland, der Schweiz, in Frankreich, Belgien und England. Gotha, 1863.
4. James Cattell, *American Men of Science: A Biographical Dictionary* (New York: Science Press, 1906).
5. For further analysis, see Ellen Hazelkorn, *Ranking and the Reshaping of Higher Education: The Battle for World-Class Excellence* (Basingstoke, UK: Palgrave Macmillan, 2011).
6. Wendy Espeland and Michael Sauder, *Engines of Anxiety: Academic Rankings, Reputation, and Accountability* (New York: Russell Sage Foundation, 2016).
7. Hazelkorn, *Ranking and the Reshaping*.
8. https://www.umultirank.org/.
9. Wendy Espeland and Michael Sauder, "Rankings and reactivity: how public measures recreate social worlds," *American Journal of Sociology* 113, no. 1 (July 2007): pp. 1–40, https://www.jstor.org/stable/ 10.1086/517897.
10. Ibid., 6.
11. Bryan Jones and Frank Baumgartner, *The Politics of Attention: How Government Prioritizes Problems* (Chicago: University of Chicago Press, 2005).

12. Alexander Cooley and Jack Snyder, *Ranking the World: Grading States as a Tool of Global Governance* (Cambridge, UK: Cambridge University Press, 2016).
13. Giorgio Touburg and Ruut Veenhoven, "Mental health care and average happiness: strong effect in developed nations," *Administration and Policy in Mental Health and Mental Health Services Research* 42, no. 4 (July 2015): pp. 394–404, https://www.ncbi.nlm.nih.gov/pubmed/25091049.
14. Zoltan Rihmer, Xenia Gonda, Balazs Kapitany, and Peter Dome, "Suicide in Hungary: epidemiological and clinical perspectives," *Annals of General Psychiatry* 12, no. 21 (2013), https://www.ncbi.nlm.nih.gov/pmc/ articles/PMC3698008/.
15. Betsey Stevenson and Justin Wolfers, "Economic growth and subjective well-being: reassessing the Easterlin paradox," *National Bureau of Economic Research Working Papers*, no. 14282 (2008), https://www.nber.org/ papers/w14282.
16. Joseph Nguyen, "What are the benefits of credit ratings?", Investopedia, March 6, 2018, https://www.investopedia.com/ask/answers/09/benefits-of-credit-ratings.asp.
17. Denise Finney, "A brief history of credit rating agencies," Investopedia, June 4, 2018, http://www.investopedia.com/articles/bonds/09/history-credit-rating-agencies.asp.
18. "Who rates the credit rating agencies?", Quora, accessed February 10, 2019, https://www.quora.com/Who-rates-the-credit-rating-agencies.
19. Michael Lewis, *The Big Short: Inside the Doomsday Machine* (New York: W. W. Norton & Company, 2011).
20. Patrick Bolton, Xavier Freixas, and Joel Shapiro, "The credit ratings game," *National Bureau of Economic Research Working Papers*, no. 14712 (February 2009), https://www.nber.org/papers/w14712.
21. Alice M. Rivlin and John B. Soroushian, "Credit rating agency reform is incomplete," Brookings Institution, March 6, 2017, https://www.brookings.edu/research/credit-rating-agency-reform-is-incomplete/.
22. Aaron Klein, "No, Dodd-Frank was neither repealed nor gutted. Here's what really happened," Brookings Institution, May 25, 2018, https://www.brookings.edu/research/no-dodd-frank-was-neither-repealed-nor-gutted-heres-what-really-happened/.
23. "The credit rating controversy," Council on Foreign Relations, February 19, 2015, https://www.cfr.org/backgrounder/credit-rating-controversy.

24. "What is grand corruption and how can we stop it?", Transparency International, September 21, 2016, https://www.transparency.org/news/feature/what_is_grand_corruption_and_how_can_we_stop_it.
25. Miriam Goldman and Lucio Picci, "Proposal for a new measure of corruption, and tests using Italian data," *Economics & Politics* 17, no. 1 (March 2005): pp. 37–75, https://doi.org/10.1111/j.1468-0343.2005.00146.x.
26. Mihály Fazekas, István János Tóth, and Lawrence Peter King, "Anatomy of grand corruption: a composite corruption risk index based on objective data," *Corruption Research Center Budapest Working Papers*, no. 2 (September 2013), http://dx.doi.org/10.2139/ssrn.2331980.
27. Chikizie Omeje, "Who influenced Nigeria's ranking in TI's corruption perceptions index 2017?", International Center for Investigative Reporting, February 27, 2018, https://www.icirnigeria.org/data-who-influenced-nigerias-ranking-in-tis-corruption-perceptions-index-2017/.
28. Mlada Bukovansky, "Corruption rankings: constructing and contesting the global anti-corruption agenda," in *Ranking the World. Grading States as a Tool of Global Governance*, edited by Alexander Cooley and Jack Snyder (Cambridge, UK: Cambridge University Press 2016).
29. Staffan Andersson and Paul M. Heywood, "The politics of perception: use and abuse of Transparency International's approach to measuring corruption," *Political Studies* 57, no. 4 (December 2009): pp. 746–767, https://doi.org/10.1111/j.1467-9248.2008.00758.x.
30. "Freedom in the world 2018: methodology," Freedom House, accessed February 10, 2019, https://freedomhouse.org/report/methodology-freedom-world-2018.
31. Ibid.
32. Bjørn Høyland, Karl Moene, and Fredrik Willumsen, "The tyranny of international index rankings," *Journal of Development Economics* 97, no. 1 (January 2012): pp. 1–14, https://doi.org/10.1016/j.jdeveco.2011.01.007.

Chapter 7

1. See also the movie *Bad Reputation*.
2. See the excellent book Gloria Origgi, *Reputation: What It Is and Why It Matters* (Princeton, NJ: Princeton University Press, 2018).
3. https:// www.facebook.com/ officialbaddiewinkle/ photos/ a.763814027069824/1010775902373634 /?type=3&theater.
4. Robert Axelrod, *The Evolution of Cooperation* (New York: Basic Books, 1984).

5. Martin A. Nowak and Karl Sigmund, "Evolution of indirect reciprocity," *Nature* 437, (October 2005): pp. 1291–1298, https://doi.org/10.1038/nature04131.
6. Eszter Hargittai, "Confronting the Myth of the 'Digital Native,'" *Chronicle of Higher Education*, April 21, 2014, https://www.chronicle.com/article/Confronting-the-Myth-of-the/145949.
7. Susan Gunelius, "10 ways to successfully build your online reputation," *Forbes*, December 13, 2010, https://www.forbes.com/sites/work-in-progress/2010/12/13/10-ways-to-successfully-build-your-online-reputation/.
8. "What is the digital reputation?", Social Digital Mentors, accessed February 10, 2019, http://www.social-digital-mentors.eu/index.php/4-what-is-the-digital-reputation.
9. Elliott W. Montroll and Michael F. Shlesinger, "On 1/f noise and other distributions with long tails," *Proceedings of the National Academy of Sciences of the United States of America* 79, no. 10 (May 1982): pp. 3380–3383, https://www.jstor.org/stable/12420.
10. Sidney Redner, "Citation statistics from 110 years of *Physical Review*," *Physics Today* 58, no. 6 (June 2005): p. 49, https://physicstoday.scitation.org/doi/10.1063/1.1996475.
11. Eugene Garfield, "Citation indexes for science: a new dimension in documentation through association of ideas," *Science* 122, no. 3159 (July 1955): pp. 108–111, http://science.sciencemag.org/content/122/3159/108.
12. Amy Qin, "Fraud scandals sap China's dream of becoming a science superpower," *New York Times*, October 13, 2017, https://www.nytimes.com/2017/10/13/world/asia/china-science-fraud-scandals.html.
13. Ben Martin, "Editors' JIF-boosting stratagems: which are appropriate and which not?", *Research Policy* 45, no. 1 (February 2016): pp. 1–7, https://doi.org/10.1016/j.respol.2015.09.001.
14. Wenya Huang, Peiling Wang, and Qiang Wu, "A correlation comparison between Altmetric attention scores and citations for six PLoS journals," *PLoS ONE* 13, no. 4 (April 2018), https://doi.org/10.1371/journal.pone.0194962.
15. For the top 100 articles in 2017, see https://www.altmetric.com/top100/2017/.
16. Andras Schubert and Gabór Schubert, "All along the h-index-related literature: a guided tour," in *Springer Handbook of Science and Technology Indicators*, edited by Wolfgang Glänzel, Henk F. Moed, Ulrich Schmoch, and Mike Thelwall (Berlin: Springer, 2018).

17. Malcolm Gladwell, *Outliers: The Story of Success* (New York: Little, Brown and Company, 2008).
18. Albert-László Barabási, *The Formula: The Universal Laws of Success* (New York: Little, Brown and Company, 2018).
19. Dashun Wang, Chaoming Song, and Albert-László Barabási, "Quantifying long-term scientific impact", *Science* 342, no. 6154 (October 2013): pp. 127–132, http://science.sciencemag.org/content/342/6154/127.
20. George D. Birkhoff, *Aesthetic Measure* (Cambridge, MA: Harvard University Press, 1933).
21. David W. Galenson and Robert Jenson, "Careers and canvases: the rise of the market for modern art in the nineteenth century," *National Bureau of Economic Research Working Papers*, no. 9123 (August 2002), https://www.nber.org/papers/w9123.
22. Susan Stamberg, "Durand-Ruel: the art dealer who liked Impressionists before they were cool", *NPR*, August 18, 2015, https://www.npr.org/2015/08/18/427190686/durand-ruel-the-art-dealer-who-liked-impressionists-before-they-were-cool.
23. Checked on October 28, 2018.
24. Federico Etro and Elena Stepanova, "Power-laws in art," *Physica A: Statistical Mechanics and its Applications* 506 (September 2018): pp. 217–220, https://doi.org/10.1016/j.physa.2018.04.057.
25. Jens Beckert and Jörg Rössel, "Art and prices: reputation as a mechanism for reducing uncertainty in the art market," *Kolner Zeitschrift fur Soziologie und Sozialpsychologie* 56, no. 1 (2004): pp. 32–50.
26. Alessia Zorloni, *The Economics of Contemporary Art: Markets, Strategies and Stardom* (Berlin: Springer, 2013).
27. Cooper Smith, "Facebook users are uploading 350 million new photos each day," Business Insider, September 18, 2013, https://www.businessinsider.com/facebook-350-million-photos-each-day-2013-9.
28. "Art history is exhibition history: the story behind ArtFacts.Net," ArtFacts, September 14, 2018, https://blog.artfacts.net/art-history-is-exhibition-history-the-story-behind-artfacts-net/.
29. "Updated artist ranking for 2014 now online," ArtFacts, April 5, 2014, https://artfacts.net/news/7738.
30. "Kader Attia," The Falmouth Convention, accessed February 10, 2019, http://thefalmouthconvention.com/speakers-3/kader-attia.

31. Alex Lopez-Ortiz, "Why is there no Nobel in mathematics?", University of Waterloo Math FAQ, February 23, 1998, http://www.cs.uwaterloo.ca/~alopez-o/math-faq/node50.html.
32. Andrew Brown, "The ugly scandal that cancelled the Nobel prize," *The Guardian*, July 17, 2018, https://www.theguardian.com/news/2018/jul/17/the-ugly-scandal-that-cancelled-the-nobel-prize-in-literature.
33. Robert Epstein and Ronald E. Robertson, "The search engine manipulation effect (SEME) and its possible impact on the outcomes of elections," *Proceedings of the National Academy of Sciences of the United States of America* 112, no. 33 (August 2015): pp. E4512–E4521, https://doi.org/10.1073/pnas.1419828112.
34. Robert Epstein, Ronald E. Robertson, David Lazer, and Christo Wilson, "Suppressing the search engine manipulation effect (SEME)," *Proceedings of the ACM on Human–Computer Interaction* 1, no. 42 (November 2017): pp. 1–22, https://cbw.sh/static/pdf/epstein-2017-pacmhci.pdf.

Chapter 8

1. Stacy Wood, "Generation Z as consumers: trends and innovation," *Institute for Emerging Issues* (2013), https://iei.ncsu.edu/wp-content/uploads/2013/01/GenZConsumers.pdf.
2. Carlos Gomez-Uribe and Neil Hunt, "The Netflix recommender system: algorithms, business value, and innovation," *ACM Transactions on Management Information Systems* 6, no. 4 (January 2016), https://dl.acm.org/citation.cfm?id=2843948.
3. Josef Adalian, "Inside the binge factory," *New York Magazine*, June 11, 2018, https://www.vulture.com/2018/06/how-netflix-swallowed-tv-industry.html.
4. John Ferguson, "Hellish homeless hostel exposed by *Daily Record* is condemned as worse than 'Soviet gulag' in Holyrood," *Daily Record*, December 17, 2014, https://www.dailyrecord.co.uk/news/scottish-news/hellish-homeless-hostel-exposed-daily-4829723.
5. https://www.eharmony.com/success/stories/.
6. "What is the best matching algorithm for dating?", Quora, accessed February 10, 2019, https://www.quora.com/What-is-the-best-matching-algorithm-for-dating.

Chapter 9

1. Jonathan Swift quoted in Alan Krueger, "Economists try to explain why bubbles happen," *New York Times*, April 28, 2005, https://www.nytimes.com/2005/04/28/business/economists-try-to-explain-why-bubbles-happen.html.
2. Cathy O'Neil, *Weapons of Math Destruction* (New York: Penguin Random House, 2016).

Index

Note: Page numbers followed by *f* or *t* denote figures and tables.
For the benefit of digital users, indexed terms that span two pages (e.g., 52–53) may, on occasion, appear on only one of those pages.